THE POSTDIGITAL MEMBRANE

The Postdigital Membrane

Imagination, Technology and Desire

Robert Pepperell and Michael Punt

intellect™
Bristol, UK
Portland, OR, USA

First Published in Hardback in 2000 in Great Britain by
Intellect Books, PO Box 862, Bristol BS99 1DE, UK

First Published in USA in 2000 by
Intellect Books, ISBS, 5804 N.E. Hassalo St, Portland, Oregon 97213-3644, USA

Consulting Editor: Masoud Yazdani
Copy Editor: Peter Young

A catalogue record for this book is available from the British Library

ISBN 1-84150-042-9

Printed and bound in Great Britain by Cromwell Press, Wiltshire

For Daisy

For Daisy

Contents

Acknowledgments   1

Introduction   2

Section One   7

Section Two   27

Section Three   40

Section Four   55

Section Five   71

Section Six   87

Section Seven   107

Section Eight   125

Section Nine   143

Section Ten   157

Notes   167

Bibliography   177

Acknowledgments

Books such as this emerge from a complex network of influences, desires and particular circumstances that allow them to be fully realised. Certainly without the support of our publishers, Intellect, we would not have been spurred to commit our extended discussions to paper with quite the determination that a successfully completed collaboration requires. Aside from a platform to express ideas, research demands time and an intellectually energising environment. In particular we would like to thank the Principal of University College of Wales, Newport, Professor Ken Overshott and Professor Roy Ascott, the director of CAiiA-STAR for the support that they have given to this project. All books make demands on those people who are personally and intellectually close to the authors and we have been fortunate recipients of great kindness and understanding. Of necessity these contributions go unsung since, to include them all with full justice would occupy excessive space. However, it would be improper not to mention Alun Owen for his help with bibliographic research and Jane Plüer for her thoughtful advice in the final stages of the design.

During the later part of this project Daisy was born to Ruth and Robert Pepperell. As we worked on ideas about consciousness it was insightful to watch her mind grow as she began to be conscious of the world. Since she offered such a unique combination of developing intelligence, pleasure and distraction at a crucial moment, and since she will certainly emerge from the postdigital membrane as the kind of human being that none can hope (or would want) to fully predict, we thought it only proper to acknowledge her. In doing so Daisy will, in the inclusive spirit that pervades the argument that follows, stand in for all those other humans, without whom, this book would not have been written.

Robert Pepperell and Michael Punt

December 2000

Introduction

This book is about how imagination, technology and human desire are understood on the cusp of a new century — a century that is being defined in terms of the 'digital-age'. Far from subscribing to this general diagnosis, however, we argue that the intellectual restrictions of the digital paradigm are now becoming unavoidable, not least since it insists on the reduction of continuous reality into discrete binary units. Our position in this book demands that whilst the continuum of reality should not be sacrificed for the sake of conceptual convenience neither should we deny the significance of the digitisation of information. The term 'Postdigital' is intended to acknowledge the current state of technology whilst rejecting the implied conceptual shift of the 'digital revolution' — a shift apparently as abrupt as the 'on/off', 'zero/one' logic of the machines now pervading our daily lives. New conceptual models are required to describe the continuity between art, computing, philosophy and science that avoid binarism, determinism or reductionism. The very unpredictability and ambiguity of human experience — its most valuable features — are being reconciled in the binary codes of digital processing and the logical prescriptions of many scientists. These amputated descriptions expose the need for more flexible metaphors with which to describe the stable yet dynamic reality of the postdigital age. The metaphor we use in this book is that of a biological membrane, a lubricating sheath that gives form to complex phenomena (such as imagination, technology and desire) at the same time as enabling a continuity between them. The power of the membrane metaphor is its dual and contradictory function: like a transparent wall, it both connects and divides. If we are to talk about an environment that is both changing and staying the same, or things that are both separate and integrated, we need mental tools more subtle than exclusive, binary logic can supply. As a consequence, we insist that formal analysis of human culture is inadequate. Cutting up the data ever more finely, we assert, will only obscure the thickness of the membrane by reducing it to a thin film that will, at best, merely reflect the kind of scalpel used and, at worst, reiterate current intellectual vogues. In its place, we claim that the membrane must be described through both analysis and sensation, and that this description must never be confused with explanation. However the reader chooses to approach this book — sequentially, rhapsodically, with antagonism or generosity — the authors hope its reception will expose the necessity for a new kind of diverse order.

The impulse for us to collaborate on this book arose from independent research that we both conducted in the 1990s along parallel tracks. These resulted in two monographs, which were quite specific to the decade, and appeared to demand reconciliation. In 1995 Robert Pepperell published *The Post-Human Condition*. This was a polemical manifesto that discussed how we saw ourselves, and the world, in light of the recent dramatic increase in co-operation between humans and technology. One prominent thesis of this book was that reality, which was formerly understood as the counterpoint of imagination, was increasingly seen as continuous with it. Accounting for this, he argued that machines which were thought to replicate human intelligence, whilst grounded in a dubious logic, did indeed shift the perception of what it meant to be human. At the same time Michael Punt was conducting research into nineteenth century technology (in particular the Cinématographe) in order to answer a number of fundamental questions about the processes of invention and technological innovation. Punt's conclusions, published in *Early Cinema and the Technological Imaginary* were that doctrinaire histories of technology were asymmetrical in that they failed to factor in the users as active interpreters of technology — users who were able to change the meaning of an invention through a process of mutual intelligibility. As a consequence, histories of technology could not account for the discrepancy between what inventors thought they were doing and what they apparently achieved. Extrapolating from this insight with case studies from the late twentieth century, it became evident that to understand the effects of the human-system collaboration (which Pepperell had posited) it was necessary to re-think what being human meant.

As common ground emerged between the two studies, the subtitle quickly declared itself — imagination, technology and desire. This uncertain triad stands in contrast to the robust triad of social, economic and technological explanations of historical change. The author's scepticism about such determinism is founded on a belief in the importance of the body, and the inherent failure of our attempts to fully satisfy

3

its desires. Crucial amongst all desires is the need to extract order from incoherent experience. One observable symptom of this condition, we argue, is that historical change is often seen as essentially linear with abrupt punctuation, whereas, in reality, transitions between states are attenuated and messy, and can only be accounted for with hindsight. Punt's case study of the emergence of the cinema noted the contrast between 'goal-driven' histories of invention as a relay race toward realism — moving pictures, sound, colour — and the archival evidence showing that these technological possibilities existed well before the cinema was instantaneously 'invented'. Factoring in the user's imagination and desire into technological change, however, helped us to understand why a particular entertainment form emerged when it did. The view of historical change that both Pepperell's and Punt's work advocated was not posited on cause and effect, nor was it dialectical. Instead a 'messier' model was necessary in which contradictory forces were neither resolved, nor neutralised. They simply co-existed as identifiable floating forces, accommodating and generating human desire.

The task of the cultural analyst and the historian in explaining historical change is to track as much of the visible network of forces as possible, and evaluate their relative determining power in relation to verifiable outcomes. In this methodology, the best the analyst and historian can hope to achieve is a useful description of the 'mess' — never a conclusive explanation of the moment. It was this shared idea of how things change, and have changed, that shaped the title 'The Postdigital Membrane' and determined the rhetorical strategy of the thesis. The argument in this book, expressed in text and images, is not without difficulty for both the authors and the readers. These difficulties are unavoidable since we cannot employ the chronological templates of the academic historian, nor the ideological tent-poles of the cultural critic. Instead, we have the singular certainty of an eternal present, a multi-dimensional instant that voids and consumes what others think of as the 'past' and the 'future'.

At difficult moments throughout the writing process we have been haunted by the vivid image of a person trying to repair the punctured tyre of a bike they are also riding. Using a multitasking approach, we have asked what might happen to ideas that disrespect the usual authorial hierarchies or presentational structures; what might happen when claims are subject to different categories of evidence and voices shift from 'past' to 'present', from impersonal to colloquial; what new possibilities might transpire if asides and ironies are collapsed into a single picture, and fully developed arguments are presented as one-liners written with the portentous ring of an advertising copywriter? One effect of this thought process has been to shift the balance of interpretation towards the reader as they engage with the thick membrane of text and images we provide. We had imagined a table in a café where our discussion went on, where an empty place was set for the reader. This reader (the same reader present over any writer's shoulder) we imagined as a fully desiring, articulate human who was anxious to make interventions — to quarrel, agree, become exasperated and possibly even slightly bored, or anxious to reach for the fast forward button or re-edit the tape with reckless jumps. We do not expect that the whole book will be essential reading for all readers, nor do we expect that the chapters will be read in sequence (Who does that these days anyway? Who ever did?) So if you happen to be scanning this in a bookshop then perhaps that is how we meant it to be encountered — standing, slightly distracted, and uncertain. The distinct pleasures to be had from browsing, eaves-dropping in cafés and over telephones however, precipitate the kind of fascination that only partial data can bring. We hope that this book offers these distinct pleasures as a consequence of its rhetorical strategy — to suggest more than we say.

This book is divided into ten sections, each beginning with a formative assertion or proposition which serves as a useful springboard for discussion. Each section consists of images and words that embellish, and sometimes deviate from, the initial premise. The text contains of a number of sub-titled paragraphs which, as we have said of the sections themselves, can be approached in almost any order. Many of the images are included not to illustrate specific pieces of text, but as suggestions of ideas that cannot adequately be expressed in words.

Side by side the authors shared their hard-disks, exchanged text by email, discussed ideas by telephone and took digital cameras and scanners to the world to make images in parallel with the conversations and writing. The result was the ideas contained herein. We offer the reader these ideas with a deliberate directness and economy which, we feel, is often missing from academic works.

Since we do not claim that the postdigital membrane is a complete theory, or even a coherent set of ideas, we wish to leave open as many doors as we try to close. To this end, in parallel to this book, a web-site has been established at 'postdigital.org' where we intend that the debate will continue among those who occupy the empty chair at this discussion.

Section One

PROPOSITION:
Technology is the tangible expression of desire motivating human imagination to modify reality.

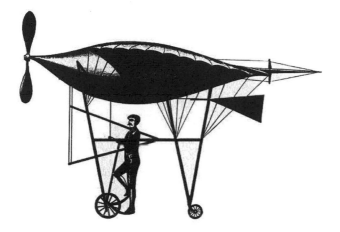

Why do we have technology?

Surveying history, it seems that there are at least two, often closely connected, reasons why humans develop technology. One is to accomplish some fantasised wish such as recording sound and light. Another is to decrease the amount of time and/or effort required to achieve some desirable end, such as carrying coal from one town to another in order make life easier or to increase profit. Profit is seen as desirable in itself as it leads to the accumulation of wealth, and wealth can be used to realise fantasies or reduce the amount of time and/or effort required in carrying out the tasks of life. Whilst there are many other explanations for technological change (such as extending human mental and physical abilities or overcoming the arduous burdens of nature by making things more accurate, safer, cleaner, healthier, etc.) it is consistently the case that technology has a nominated purpose — the attempt to satisfy some human desire.

Technology removes obstacles to desires

Put another way, technology removes obstacles to the satisfaction of desire. At an early age we apparently develop a sense that the world resists or constrains our behaviour and we are often denied what we want. This central aspect of the human condition can be circumvented to some extent by adapting our behaviour or surroundings so as to meet with our desires. When we want to crack open a nutshell to eat a nut and the shell is too tough we can use a rock or stick to break it. If we want to draw up water from a well with less effort then we might devise a mechanism for doing so. If we want to make yarn quicker than a hand spinner we can invent a machine to do it. These kinds of technological intervention modify aspects of reality according to our needs, in spite of the various obstacles that appear to constrain our desires. In this reciprocity of desire and technology we are always trying to resist the entropic tendency of the Universe and get more for less, without any apparent limit to how much more we want.

Imagination, technology and desire

What we still want

In spite of (or, perhaps, because of) our considerable technical success, there are many human desires and aspirations that still resist realisation; the desire to live indefinitely, to travel through time, to have cost-free energy sources, to make contact with alien life-forms, to read other peoples minds, to create artificial beings, and so on. The fact that we have not yet devised the technical means of realising these goals does not stop us from imagining that we might in the future. Indeed, this list of phantasmagorical human dreams may be driving some of the most prestigious and highly funded scientific research programs in history — The Human Genome Project, High End Particle Acceleration, Cold Fusion, SETI (Search for Extra-Terrestrial Intelligence), Brain Imaging and Intelligent Robotics, all of which feed an imaginary, pre-Oedipal future in which desires are fulfilled.

Why do we transform the world and ourselves?

Changing the operation of reality is not always easy. The fact that the world generally resists us, but occasionally yields to our ingenuity, encourages the imagination of the designer, the inventor, the programmer, the artist, the songwriter, the scientist or the writer to continue their exploratory struggles. Despite the obvious difficulties inherent in any creative practice, in which a high ratio of unproductive effort combines with a great probability of failure, there is an assumption that success is possible. We subscribe to the notion that the ground may 'give' at any time and a solution to an intractable obstacle to desire might be overcome. Perhaps the real questions historians of technological change should ask is not 'why do technologies change?' but 'why do we invent?' Why do we continually imagine that we can satisfy desire through strategic action?

The imaginative imperative

The human imagination can see things that are not there, which is both a gift and a curse. It is a gift because it allows us to escape from harsh reality into congenial fantasy and a curse because it can create dissatisfaction with the present by projecting the possibility of something better than we have. The white elephant of our imagination can summon up ideal conditions, generating utopian possibilities we try to realise, but can also lead us to fantasise the most chilling horrors that might befall us. As the body of science fiction literature clearly demonstrates, we are trying to keep up with technological developments we have already been told to expect whilst, simultaneously, living with nightmare expectations of their future failure. Cheap air-travel, disease-resistant crops and interstellar communication are all now realities where once they were dreams. But they also become the stuff of disaster stories, activist resistance and Hollywood plots.

"If the software doesn't meet the spec the spec is wrong"

Technological development only seems strategic with hindsight. Engineers, and more recently computer programmers, often solve problems by deliberately changing the underlying conception of a device. They call this kind of thing a kluge (from the German, *klug*: clever). Some complex technologies such as the personal computer are a bricollage of kluges; technologies that initially meant one thing are reinterpreted to mean another. On the Personal Computer, the typewriter keyboard does not become the impersonal intermediate stage between hand and print, but the interface for any number of operations that personalise the textual utterance. Similarly, television technology devised to satisfy the desire for tele-presence becomes a localised output device in the screen. These new meanings are brought together by contingency to produce a device that satisfies a completely different set of human demands, while retaining an occult shadow of former imperatives. It should not be surprising, therefore, that within three decades, the humdrum calculating device becomes the dominant means of global interaction. The ductility of technological meanings echoes the malleability of the human imagination, and the persistence of desire subject to Universal Laws. (see note 1).

Myths of order from confusion

Given this confluence of past and present imperatives, one begins to suspect technological artefacts might function in a similar way to historical myths in that they serve as rationalising models for those cultures that produce them. The confusion and complexity of actual events is reduced as we consider only those pieces of information that are accountable. Present realities (artefacts or social conditions) are regarded, retrospectively, as the inevitable outcome of an imagined past. In which case, technology's culturally determining role is not only in the feats of data processing or earth-moving that it helps us achieve but also in the ideas it generates about itself, and us.

10

Speaking as the Machine demands

Some time ago somebody asked a group of scholars in an internet news group (note we say scholars) what John Travolta was reading in *Pulp Fiction*; these are some of the replies:

'From my hazy recollection, Vince was reading Modesty Blaise while sitting on the toilet and in another scene too I think.'

There were a number of other similar answers:

Vincent Vega (John Travolta) reads _Modesty Blaise_ when he's sitting on the John, both times. Here is the bibliographic info: AUTHOR: O'Donnell, Peter, 1921-, TITLE: Modesty Blaise, EDITION: [Book Club ed.] PLACE: Garden City, N.Y. : PUBLISHER: Doubleday, YEAR: 1965, PUB TYPE: Book FORMAT: 252 p. ; 22 cm.

There was a 1966 British movie of Modesty Blaise directed by Joseph Losey. PULP FICTION is a celebration of pop, and MB was at the tops of the pop, so to speak. Since there are so many nods in PF to various popular culture and film I wonder if this is just one more. Does anyone else see more to it? Vince is recently returned from three years in Europe, where Modesty Blaise was widely popular as both a book and a comic strip.

Since someone asked, I've got some thoughts about the 'Modesty Blaise' book in Vincent's hands. Her picture on the cover shows a brunette pageboy hairstyle similar to Mia's, to Butch's girlfriend and the taxi driver. All but Mia also have strong connotations of foreign femme-ness. In the case of the girlfriend, there is the amazing shot of her superimposed on the war film on TV, with an enemy (Vietnamese?) soldier pointing his gun right at her. This evokes Butch's father's failure in Vietnam, and maybe Butch's desire to perform a heroic rescue and right that failure (which he achieves, with a samurai sword yet, soon enough).

Regarding PULP FICTION, I'm not positive but I think the John Travolta character was reading 'Being and Nothingness', a huge treatise on Existentialism by Jean-Paul Sartre.

11

A number of things stand out from these responses. First they are all from film scholars in respectable departments (not *Pulp Fiction* freaks or cult-movie train-spotters). Second, the usual restrained tones of scholarship and academia have been overtaken by the energetic, vital prose style and creative speculation which is the essence of electronic communications. Finally, these scholars are well trained to answer to their questions definitively by looking at the script, or even emailing Quentin Tarentino. In fact, one would think that it is a matter of professional pride that they should. There are at least two reasons why this did not occur to anyone. In the first place the dubious authority of the author as the final arbiter of meaning is so much part of modern academia that even in the case of the Tarentino — the paradigmatic auteur of the mid-nineties — his evidence is thought to be no more valuable than the reader's (even when it obviously is). But more important perhaps is that the point of the question was not primarily to obtain a definitive answer but to reinvigorate the movie after it was over. Not only does email technology spawn a new kind of discourse, it also propagates extended loops of desire that amplify the shared pleasure of the movie.

Humans tend toward resistance to entropy

Generating dissonant patterns of interferenc

We can start to appreciate that something extremely complex has to be factored into the question of desire and its relation to technology. Many bulletin boards and newsgroups (such as the one cited above) are electronic hybrids of a number of communication genres most favoured by so-called 'subcultures'. Using a potent mix of rhetorical styles, contributors to the ubiquitous conspiracy theory boards, for example, routinely speculate on events as wide ranging as the death of JFK, Princess Diana and contaminated breakfast cereal. Although conspiracy theory sites might not be the place to discover truths about any particular incidents, they are rich terrain for the cultural analyst. Their vigorous prose and imaginative invention perfectly express the mood reflected in many recent cinema releases. According to Lavery, Hague and Cartwright, the responsibility for this mood lies with what is perceived as a loss of psychological control induced by the digitisation of bureaucratic power and the efficiency of the news media. Baroque narratives, which invest the extraterrestrial and supernatural with credibility, they argue, are attempts regain some control of unexplained events by giving order to the chaos of information. That some appear on electronic bulletin boards, or in chat rooms, is paradoxical since nowhere is this chaos more apparent than on the internet. Nonetheless, the megabytes of text on the conspiracy theory boards creatively weave logical explanations using newspaper reports, forensic evidence and imagination fed by movie plots. What becomes apparent is the dissonant interference patterns generated by desire, imagination and technology as they relentlessly play upon each other.

Harmonising imaginations

The use of the bulletin board to make sense of contradictory ideas goes some way to explaining why, throughout the recent history of technological change, the dominance of one particular device over others is not solely decided by technical efficiency. More often than not it is a negotiation between the parties involved over whose mechanical representation most closely matches a shared mental model, or which permits a shared model to emerge. Generally, technological solutions that are innovated as products are those which match the imagination of both the engineer and the client. This process is dependent on all parties sharing a similar perception of the world as it currently is, and transferring some technologically modified version of this perception into a harmonious image of the future. In short, all new technologies acquire mutual intelligibility in the process of being adopted. In a telling contemporary example, Eugene Ferguson in *Engineering in the Mind's Eye* points out that the Pentagon's agreement to invest in manned space flight first required that senior figures in the government had a shared idea of what might be the eventual outcome. According to Ferguson, to achieve this, Eisenhower screened Disney cartoons to the military High Command.

We have to know what something is before we can invent it

Indulging one's desires in the hope of satisfying them is like trying to extinguish a fire by pouring petrol over it.

The counterpoint between human behaviour and technological artefact produces a further complex resonance. Almost simultaneously with satisfying human wants, technology acts to stimulate them by encouraging additional desires. We would not want a VCR if such machines did not exist just as we do not want things now which we will probably buy at some time in the future when they are invented. Because we have washing machines, cookers and central heating we can wash our clothes, prepare our food and heat ourselves more quickly and with less effort than if we did not have them. As a consequence, we have more time and energy left over to do more things and we want more things to fill that spare time. As we get used to doing things with less expenditure of time and effort we ask why other things that require time and effort can't be done more efficiently. For example, why should we have to going shopping for food or go out to rent videos or to see a film at the cinema — couldn't these chores also be automated? One could summarise this state of affairs by saying 'more things demand more things'. To give a useful economic example: because it is possible to produce yarn mechanically at 100 times the rate of a manual spinner the general demand for yarn rises as the cost per yard falls and new products and design features (like t-shirts and extravagant pleats) are devised which consume cheaper yarn. Needs that never existed before are created as a consequence of our imaginative interaction with inventions which, in themselves, stimulate our insatiable capacity for more things. What permutation of human needs were served by the introduction of 'Fluffy Bunny' slippers, mock-Tudor windows or moustache wax?

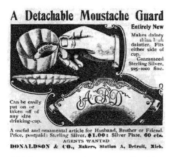

Will we ever be satisfied?

Out of time

Such is the momentum generated by the general consumption of products that even objects for which there is no apparent demand seem to emerge. The transition from invention to innovation (from prototype to product) is not an immediate but a lengthy, haphazard process in which the original meaning of the invention cannot be guaranteed. Certainly in the nineteenth century, new technologies met the world more like frail damp butterflies emerging from cocoons vulnerable to environment and predator rather than a charging locomotive. On average, it took around eleven years for an invention to acquire a consensual public meaning. In the case of the Gillette razor it was 43 years before the idea of a safe shave seemed to be a good one and the safety razor caught on. Many inventions were patented without obvious application, to lay un-exploited only to be reinvented sometime later. In 1865, for example, Alexander Parkes launched a flexible transparent material using cellulose that he called *Parkesine*. It had no significant use and was lost only to resurface thirty years later, this time being called Celluloid — the material which most histories claim the world was waiting for before moving image photography itself could be invented. This suggests that, in the story of technology, the interleaving between demand and supply is erratic rather than, as is often implied by orderly accounts, symmetrical.

Technological meanings are unstable

Sometimes, as social constructivists argue, new kinds of clients and new imperatives alter the meaning of a solution from that which the inventor envisaged. The story goes that after some Earthly catastrophe in the year 2020 a group of alien scientists from Mars came to earth and sifted through the remains to find out what they could about our civilisation. They found among other things a movie projector and some film. Back on Mars after some deft reverse engineering they deciphered the logic of the technology and got the machine to work. The great and the good assembled for the first screening to get some idea about us and our culture so that they could make sense of other archaeological fragments they had salvaged from the ruins. Needless to say that when the film was projected it was not Thomas Edison but Donald Duck who quacked 'hello!' And needless to say the Martians had difficulty thereafter reconciling the remains of a Cadillac with a two-foot duck. What the Martians lacked, in this story, was an imaginative response to the apparatus. They assumed a direct correlation between the design of the machine and what it was used for. Machines are not just technology, they are repositories of meaning. As Lucy Suchman has pointed out "Every human tool relies upon, and reifies, some underlying conception of the activity that it is designed to support." These underlying conceptions are not fixed; they are subject to both historical and geographical contingency. Ships' canon become mooring bollards, fire extinguishers are often nothing more than elaborate doorstops. Working at the 'front end' of the technology, that is where a machine meets culture, often produces meanings which had no place within the scheme of the inventor's desire

16

Desire: there's always something missing

What if technology removed all constraint on our desire?

In spite of the frequently unsatisfactory results we keep on and on inventing. Is there no limit to the economy of imagination, technology and desire, or will there be a technological 'glass-ceiling' below which our desires will hang like so much evening mist? We could speculate that tools and techniques might emerge to meet ever more subtle demands that, in turn, will stimulate unknown desires. If we are to continue along recent economic and technological trajectories, there will be more things to want that require even less effort to operate. Perhaps this will continue indefinitely — technology expanding infinitely along with our desires? But technical limitations are not the only constraints on satisfying our insatiable wants. As Freud claimed, we are just as tightly enclosed by social pressure — the learnt patterns of response and self-control that allow us to integrate into our cultural mores (what he broadly termed *repression*). These restraints also determine that many of our desires remain as fantasies: wishes that often find expression as technological dreams. We know, however, these social inhibitions can be

17

transgressed by reality-simulation devices of great authenticity in which participants could meet their wishes without sanction or shame. In the 1973 Hollywood movie *Westworld* visitors to a futuristic theme park populated by cyborgs act out their historical and sexual fantasies using robots, without check or fear of reprisal. The visitor's pleasure is ultimately circumscribed, not by guilt at their own actions, but technical failure of the cyborgs, leading to the inevitable carnage.

Everything on demand

When 'Everything on Demand' becomes a deliverable technology, how could we then 'want' anything else? In his short story *The Machine Stops* (1909) E. M. Forster creates a vision in which all human needs are met by the ubiquitous, sub-terrestrial 'Machine':

> Then she generated the light, and the sight of her room, flooded with radiance and studded with electric buttons, revived her. There were buttons and switches everywhere — buttons to call for food for music, for clothing. There was the hot-bath button, by pressure of which a basin of (imitation) marble rose out of the floor, filled to the brim with a warm deodorised liquid. There was the cold-bath button. There was the button that produced literature. And there were of course the buttons by which she communicated with her friends. The room, though it contained nothing, was in touch with all that she cared for in the world.

In a scenario presaging the collapse of the Soviet social system in the 1980s, the 'Machine' purports to meet all human requirements of pleasure, transport, sustenance and knowledge, but inevitably self-destructs under the weight of its own decadence. Forster's warning against the degeneracy of artificiality is echoed in subsequent science fiction narratives and seems to expose a paradox in our attitude to future human development. On one hand, we envision a comfortable existence where all peccadilloes are attended to without the slightest cost or effort on our part. On the other hand, we achieve this only by alienating the 'natural' and 'humanistic' aspects of existence, by becoming over-dependent on machines, or indeed by becoming dehumanised. Thus, an apparent utopia is in fact a dystopia.

18

In this form of story, the anxiety arising from this paradox is usually resolved in a catastrophic realisation of the error that comes too late, as Forster makes clear:

> "But Kuno, is it true? Are there still men on the surface of the earth ? Is this — tunnel, this poisoned darkness — really not the end?"
>
> He replied, "I have seen them, spoken to them, loved them. They are hiding in the midst and the ferns until our civilization stops. Today they are the Homeless — tomorrow.."
>
> "Oh, tomorrow — some fool will start the Machine again, tomorrow."
>
> "Never," said Kuno, "never. Humanity has learnt its lesson."
>
> As he spoke, the whole city was broken like a honeycomb. An air-ship had sailed in through the vomitory into a ruined wharf. It crashed downwards, exploding as it went, rending gallery after gallery with its wings of steel. For a moment they saw the nations of the dead, and, before they joined them, scraps of the untainted sky.

The future never arrives

In contrast to many fictional accounts, the future is never some kind of streamlined utopia or dystopia in which any technology has reached absolute resolution or ubiquity. Take a time-traveller from a city street circa 1900 and transport him to the same street at the dawn of the twenty-first century. Once he has acclimatised to the 'horse-less carriages' and synthetic fabrics he would probably be more shocked by the similarities with his own time than the differences. Perhaps he would even be disappointed at the small amount of progress we have made. Just as the architectural fabric of many English County villages has remained unchanged since the eighteenth century, the view above shop-front level in many city streets is virtually identical to 1900. However, this is no more exceptional than the fact that in the postdigital age we will still wear trousers, live in brick houses and eat from wooden tables as there is no compelling reason why these accessories to existence should change. In many ways, it is because they are legacies of past desires that we value and retain them. Despite this, some combinations of contemporary desire would be less comprehensible to the time traveller confronted by John Travolta on the 'john' and IRC slang, just as something like the 'detachable moustache guard' is incomprehensible to us now. The lesson we should draw is to remember that the future will also be a 'mixed bag', much as the present is. Things will change, and stay the same, but desire will be a stranger in the land of half-familiar objects yet to be invented.

Waiting for a need to be filled

Neophobia

It is a fallacy that people welcome novelty (as many advertisers assume). We pay attention to it only when it might be a threat or an opportunity. We can embrace the new if it is a comfortable variation on an existing form, but genuine innovation is very rare (and even more rarely welcomed). The internet, for example, contains hardly any new ideas. As a means of accessing and distributing information it may be relatively new, but the information on the exponentially growing number of web-pages is itself is over-familiar, repetitious and incomplete. What is novel (and what makes it interesting) is the heat generated by the sheer quantity of illegitimate and incompatible data jostling for attention. What we like about the internet is that it gives us what we always wanted under a flag of convenience. A opportunistic technology that has caught the attention of 'old-time' speculators, panhandling the past as though it were tomorrow.

Using pleasure to regulate pleasure

Perhaps the oddest paradox of the evolving constituency of the internet is the enthusiasm with which education has embraced it, which must surely account for some of the magnificent rise in the number of users. Given that the environment is largely now commercially driven, and that the information retrieved is often trivial, inaccurate, subversive or esoteric, it is puzzling that school children are encouraged to use it as a primary source of knowledge. To use a culinary metaphor, the internet is to learning what McDonalds is to cooking. Which is not to say that fast food is not sometimes hugely entertaining, but it is pointless to give it a high profile in the kitchen. A brief reflection on media history however might explain the apparent paradox. The diffusion of popular technology has always posed something of a contradiction for the socially minded intellectual. Reformist movements at the turn of the twentieth century were just as concerned with reinstating the existing social hierarchies, which cinema threatened to overturn, as they were about the detrimental effect on morals that darkened spaces and subversive ideas had on working class audiences. The outcome was high-class moving pictures, which were rather less fun than what preceded them. Harnessing the persuasive powers of screen technology without displacing its venal attractions was the feat that successive administrators failed dismally to pull off.

Technology, like imagination, has drawbacks as well as advantages.

There is little doubt that, as regulation bites, the internet will become increasingly dismal. The regulation of cultural values was the dubious stalking horse that situated broadcasting in the classroom, first with radio and then television, despite the minimal evidence that broadcasting was effective as a teaching medium. Education also justified personal computers in the home and then the classroom, and in the process reconstructing the games machine as a corporate apparatus for writing term papers. Those technologies that start life as devices of entertainment are coerced into the service of social improvers — often with malignant consequences. For example, the web is without doubt rich source of information for specialist researchers who are able to verify the material that comes to them. But what was once a magnificently entertaining and irresponsible playground (where one could check out one's mortal expiry date) which was intended to liberate human expression now becomes an virulent surveillance and marketing tool. The lesson is that we do not necessarily gain what we are supposed to gain when we use technology to improve our lives — technology is not necessarily progressive.

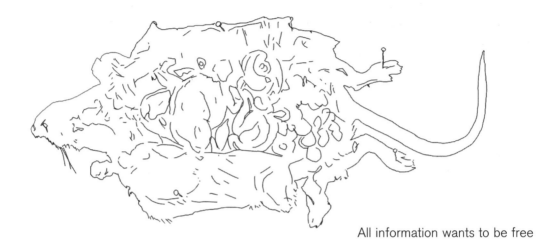

All information wants to be free

Changed by thought

What becomes apparent from consideration of the accumulated complexes and paradoxes associated with the technological is that the operation of machines cannot be reduced to the mechanical or the material. Implicated within the cogs, buttons and wires is some more abstract representation of ourselves as we would wish to be. Aside from its technical affordance as a representation, the phonograph was also a representation of a mental model that was transmuted by contact with other mental models that incorporated it as a heuristic. The business machine emerging from Thomas Edison's laboratory rapidly acquired a new meaning as demonstrators, and subsequently members of the public, used it to reduce the distance between the 'star-turn' and their distributed audiences. Similarly, the VCR — a machine intended to time-shift broadcast television — became a technology to simulate the experience of the theatrical exhibition of movies, or a source of private erotic pleasure. Beyond that, it satisfied new modes of consumption as every viewer had the potential to re-edit and rewrite a film. On the one hand this spawned new genres of action and horror movies in which spectacular effects and erotic images could be savoured repeatedly. On the other hand, a new University discipline of Film Studies garnered scholars and students who were motivated to recoup the concept of culture so as to include the experiences of ordinary people. In this mode technology is like the mythical Flying Dutchman: a mobile idea (merely represented by an object) which periodically takes on a particular material form when it meets human desire.

Technology as it is perceived

Technology may be manifest in bits of hardware, systems and bio-engineering, but such artefactual evidence should not mask its psychic dimensions. Object driven histories and aetiologies overlook the social imperatives that drive technological change as well as the role of the imagination and desire in shaping the transformation of artefacts once in the public domain. For the technological developer and entrepreneur this interference with the trajectory of their own vision can cause certain tensions which become apparent in the confusion that surrounds some new devices. All commercial developments are marketed as improvements on existing objects or processes, but functionality can actually be retrograde leading to limited demand (as was the case with Philip's proprietary CD-i machine). VHS video seriously degrades both sound and image quality rendering many highly crafted cinematic works almost unreadable; and, until recently, the personal computer merely gave us a mid-nineteenth century standard of text, picture and telegraphy. As we become caught up in this story of erratic technological development, the correlation between what we need and what we get can become corrupted. We are expected to buy ever-faster processors but few of us can input text at a greater rate than sixty words per minute, leaving 99% of the processor power dormant for most of the time. To compensate for this, software and hardware developers add more 'features' (most of which we never use) that consume the extra power, leaving us pretty much where we were before we upgraded. To judge by the optimistic tone of much marketing material, the disappointment and frustration this can cause is rarely (if ever) factored into the fantasies of the manufacturers self-image.

Technology as a membrane

We know that orthodox histories see technology as driving culture, but it is also true that culture can recast technologies, or at least the ways they are presented, according to its own demands. In deference to consumer sensibilities advertisers reclaim 'Green' credentials by promoting the eco-friendly aspects of products which malign the natural world, such as petroleum, power, and intensively farmed food products. Taken together these competing forces of marketing, invention, consumer demands and technical frailty may be thought of as a thick membrane of activity, a lubricating fluid in which the meaning of a device is decided by the complex interaction of desire, culture and contingency. Every technological artefact or system, even at the cutting edge, is a dense cultural fabric in which the threads of desire and imagination cannot be traced as separate strands, but must be seen as an diverse ecology with moments of comprehensible stability.

Technology is tangible imagination

We have argued that technology is neither autonomous nor external to human consciousness. Rather, it is better understood as an actualisation of thought that transcends our 'immaterial' consciousness. As a consequence, we must acknowledge technological artefacts as an aspect of our own imaginative desires and recognise that the things we make don't exist independently of us nor are we subject to their control, as some orthodox narratives suggest. Products of human activity are more usefully considered as tangible extensions of an imagination stirred by our desires — no product can be purposefully made without an element of desire being involved. A pen, for example, is a material embodiment of human thought and a whole range of desires as much as it is an embodiment of human labour in the economic sense. Hence, the role of imagination is not only to reflect upon the world but also to become the world as we deposit part of our being in the things we make and do. In this way our tangible thoughts become entangled with other people's thoughts when they sense what we do, perhaps in ways that sometimes we are unable to predict, control or understand.

Imagination, technology, and desire

The threads that connect technology and imagination are densely knotted and can only be un-picked with the infinite patience as ideas turn into realities and provoke other unexpected ideas. What this analysis suggests is a complicated, non-reducible matrix of histories and possibilities; a conception of technology that is intimately bound up with human imagination that is itself prompted by desire. If we try and separate out these terms and what they refer to we only end up having to put them together again, having gained little in the process. We cannot hope to extract a determinant, a final cause or to say which comes first, since even desire is constructed through the very objects designed to placate it.

The imagination is prompted by human desire to modify the world through technology, which in turn prompts desire.

Imagination, technology, and desire

The threads that connect technology and imagination are densely knotted and can only be un-picked with the infinite patience as ideas turn into realities and provoke other unexpected ideas. What this analysis suggests is a complicated, non-reducible matrix of histories and possibilities; a conception of technology that is intimately bound up with human imagination that is itself prompted by desire. If we try and separate out these terms and what they refer to we only end up having to put them together again, having gained little in the process. We cannot hope to extract a determinant, a final cause or to say which comes first, since even desire is constructed through the very objects designed to placate it.

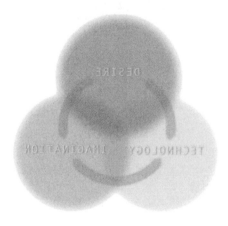

The imagination is prompted by human desire to modify the world through technology, which in turn prompts desire

Section Two

PROPOSITION:
My awareness extends to, and consists in, those things of which I am aware.

Trees also have roots in the sky

Nature is a continuous phenomenon, though we do not know in all cases how things are connected. (Illustration: Human consciousness depends on the properties of protoplasm, the existence of which depends on innumerable physical conditions peculiar to this planet; and this planet is determined by the mechanical balance of the whole universe of matter. We may then say that our consciousness is causally connected with the remotest galaxies; yet we do not know even how it arises from — or with — the molecular changes in the brain.)

Aleister Crowley, *Magick in Theory and Practice*

Thoughtful marks

When we express our imagination by embodying it in some form or act we invite the possibility that it will have an effect on others. Conscious beings may sense such expressions (words, gestures, marks, signs, and so on) and respond to them mentally and physically. In this way our own thoughts become implicated in the thoughts and actions of others through the agency of the things we make or do (the words we speak or the signs we make). Consequently, their thoughts become bound up with ours such that continuity exists between thoughts. For example, the urge to express an opinion on a public wall ('Elvis Lives') becomes embodied in paint marks. These marks can be seen by others who learn something of the opinion expressed. One might say that a significant part of the thought (or hope of Elvis's continued existence) is embodied in these marks and is transferred, via the agencies of paint and light, to other minds.

Thoughts are not confined to the mind

This general principle applied to other instances of human expression, such as the movies, computer programmes, works of art and books, challenges the unwritten assumption that thought is confined to the mind, or any one area of the body such as the brain. It suggests that the constraints of the human frame do not necessarily correspond with its limits. This is not a recklessly speculative idea; some philosophical conjecture in the field of Artificial Life has argued that inanimate objects can be regarded as self-organising systems with the potential for autonomous consciousness. Such suggestions are provocative and not without their sceptical rebuttals. But if we proceed with generosity, and accept the idea that objects and systems could have a self-awareness, then perhaps we might conclude that current conceptions of consciousness which are limited to the human brain are in themselves somewhat limited.

Beyond Embodiment

Film theorists generally agree that Classical Hollywood Cinema is best understood as a quasi-autonomous system of production. As movie making became more focused on Hollywood, scientific principles of task management, derived principally from Frederick Taylor, shaped the form of films. The studio system divided what had previously been a relatively seamless production process undertaken by hierarchical teams into discrete defined functions. The influential technology of the continuity script, which organised the individual human contributions to the production process, also shaped the dynamics of the narratives and their visualisation. This device favoured screenplays that could be written according to defined organising principles of film production, and the scientific management of these tasks is self evident in the structure of Hollywood narratives. The three-act screen-play, in which the outer two are of equal duration and the middle act is the same length as the two combined, became something of an orthodoxy in the Classical period (generally agreed as being between 1917 and 1960). Such internal mathematical logic suggested even finer symmetries between significant events. The burst tyre in the early minutes of *Sunset Boulevard*, for example, matches the timing of the gun shot that kills the hapless writer Jo Geddes shortly before the end. The mid-point of the film is precisely marked as Geddes' fate is sealed with the words "Happy New Year Darling" and he is drawn into the clutch of Norma Desmond, 'an old-time movie star'. Such sophisticated invention arises from an interaction between two organising systems: human intelligence collaborates with the intelligent system of production. This only becomes possible once the convention that the mind is restricted to the human body is relinquished.

The human mind is visible in the objects we invest it in

Compelling fictions become indices of real behaviour

The brilliance of the symbiosis between the production strategies and the conventions of film narratives was in the apparent artlessness of the system. Classical Hollywood Cinema commanded the resources of daily life with the freedom of the ideal consumer. The spectator's experience of the real world was represented in the idealist *mise-en-scène* where the effort of writing and the tyranny of production were invisible, thus cloaking the operations of ordinary life. In its institutional phase at least, Hollywood cinema had the appearance of an omnipresent mind machine. Provided that the film could maintain its realistic pretence, it was apparently able to assimilate, process and resolve almost any human drama — romance, war, betrayal. Its authority was determined by self-reflexive codes that penetrated the world and infiltrated real behaviour through continuous exposure to ideal states. Hollywood cinema was a technological system that ultimately manufactured itself and, of necessity, reduced the subject by conflating consensus with common sense. It became the effective public realisation of the scientific desire to replicate human life as intelligent and self-aware (in the kitchen-Freudian sense) without the messy inconsistencies of actual existence. This kind of cinema provides a grounded example of encoded thought outside the brain.

Where are we then?

Away from coarse nature and the raucous streets, philosophers have long tried to define the relationship between 'mind', 'body' and 'reality' — the problem Hollywood resolves so well. Philosophical thinkers can be crudely divided into two camps, although the boundaries are often indistinct. There are those known as Idealists, such as Plato, Kant and Hegel, who doubt the possibility of knowing anything beyond what our senses tell us, even doubting whether anything might exist beyond them. They are opposed by Materialists, such as Democritus, Diderot and Marx who insist on the prior existence of a real world that is external to, but reflected in, human thought. A consequence of both views was the perpetuation of an opposition between the internal (subjective) world of the mind, consciousness, thought and the external (objective) world of reality, nature and the universe (Pepperell).

30

Subjectivity and objectivity

In what is broadly called 'Western' culture the opposition between subject and object can seem so integral to our way of seeing the world that, in terms of sentence construction, it underlies the very grammar of language. We take the subject to be the 'doer' of a verb and the object to be the thing to which it is done. Although in Medieval philosophy this meaning was reversed the basic distinction remained. This grammatical opposition has become embossed at the conceptual level of thought, reinforcing the split between subjective and objective reality — between me, you, he, she, them and us. This is not true of all human culture.

> The object is an object for the subject,
> The subject is a subject for the object;
> Know that the relativity of the two
> Rests ultimately on one emptiness.

Seng-ts'an, *Believing in Mind*

The 'boxed body' fallacy

The assumption of an essential separation between subject and object is so historically and culturally pervasive that it often remains unquestioned, even in revolutionary visions.

> The fundamental distinction between the materialist and the adherent of idealist philosophy consists in the fact that the materialist regards sensation, perception, idea, and the mind of man generally, as an image of objective reality. The world is the movement of this objective reality reflected by our consciousness. To the movement of ideas, perceptions, etc., there corresponds the movement of matter outside me.

Vladimir Lenin, *Materialism and Empirio-Criticism*

One would not necessarily disagree with the mirror metaphor invoked here by Lenin: that what is 'outside' him (objective reality) is formed as an 'image' in his mind through sensation, perception, etc. However, although he is careful not to suggest a lack of correspondence between reality and the 'image' of reality, he does imply that they are distinct. Lenin perpetuated what one might call the 'boxed body' model of existence, sometimes thought to accord with 'common-sense'. This assumes that the human body has a fixed boundary; the mind resides inside the walls of the box whilst the reality upon which mind reflects lies outside. The box of course is perforated to allow sensations to flow in and waste to flow out. But, according to this model, the inside and the outside of the box are discrete and retain their own qualities. One then argues about the relationship between the two sides — can one really know what's outside the box from the limited data we receive? Can we confirm anything other than our own existence? — and so on. The model proposed here, by contrast, does not prescribe limits to the human being; the conceptual box is removed. Therefore, there is no 'inside' or 'outside' to distinguish, except in anything other than an arbitrary or contingent sense. This postdigital model accepts that we mentally create distinctions as we try to understand the world, since consciousness arises through the fragmentation of reality. The mistake is to equate these imagined distinctions with an external reality. Such imagined distinctions are still real (since they have a real existence in our minds) but do not pertain prior to us becoming conscious of them.

The mirror as mind

The mirror as mind

This metaphor of the 'mind as mirror' which pervades much philosophical thought is persuasive inasmuch it confirms our preconceptions about what a mirror does — that it creates a virtual image of what surrounds it without itself being a component of those surroundings. If it were a component it would (logically) have to reflect itself and this would be apparently absurd and infinitely regressive. In other words, it is a classical observer. Hence we are able to separate the reflected-virtual image from the real world it reflects, just as we tend to separate mental reality from external reality. But this polarised conception of the mirror betrays a limited appreciation of what a mirror normally does, which is to reflect light coming from surfaces that also reflect light. Inevitably, the mirror deflects light back directly to the reflective surfaces in front of it so that this light, in turn, is deflected back to the mirror, ad infinitum. Therefore, what appears to be a bi-polar reflexive act is an infinite accumulation of reflections. The reflective surface intervenes in the world that it purports to objectively observe. More generally, it could be argued that what one sees as the 'objects in themselves' is actually the accumulation of reflections and deflections caused by the interaction of objects and light. Streams of photons reach our retina to form the consequent mental image, and so the 'real' object and the object apparent in the reflection are each presented to us as streams of reflected and deflected light. From a purely visual standpoint, there is little to distinguish the real from the reflection. However, if we believe that images of things and reflections of things are not 'the things in themselves' but only observable consequences of them, we have made an error. It would be making the same mistake to think that a photograph of a thing does not constitute the thing 'itself' but is merely an observable consequence of an optical recording. There is no intrinsic distinction between things and their consequences, nor between objective reality and its reflection in the mind.

Things are made up of their consequences (and other things)

To ground this proposition with an example: looking at your face in the mirror creates a reversed image that you can see. What is the relationship between these two faces? Are there actually two faces or is it, rather, that a single object can be

The image in a mirror is as much part of the thing reflected as the thing itself

identified by more than one observable consequence? One could justifiably argue that each face is demonstrably the same face since the reflected face, being a consequence of the real face, is also a constituent of it. To give another example in which an object must necessarily include its consequences: heat and light from the sun radiate from its (roughly) spherical surface into the infinity of the Universe in all directions at once. The sun not only consists of a ball of hydrogen 900,000 miles in diameter, but also of the infinite spread of rays that emanate from it; such rays are merely constituents of the sun that have moved away from its core. A small proportion of the rays travel the ninety three million miles to Earth, a journey that takes about eight minutes. We normally think of the sun something remote in the sky, but it clearly extends to the Earth in the form of heat, light and the radiation that surrounds and sustains us — the sun exists just as much here on Earth as it does in the sky.

Any consequence attributable to a thing is a constituent of the thing.

In a remark that would be true of anything, one could say that the sun consists of all the consequences of its existence; the sun's energy becomes a constituent of anything it interferes with (or that interferes with it). One remarkable effect of this extension to objects is things that ordinarily seem separate become continuous and, indeed, affect each other's constitution. For example, a small proportion of the rays from the sun are absorbed by plants on Earth and used to enable energetic photosynthesis. In this way, the sun affects the plants and the plants affect the sun (inasmuch as the plant interferes with that part of the sun's rays that reach it). Unequal as this interaction is, each modifies the other and becomes mutually implicated in each other's constitution. In this example, the sun is part of the plant and vice versa. That is not to say that, being a constituent, it comprises the thing in its entirety — it may be only a fractional part. Similarly, the image in the mirror, being a consequence of that scene which is reflected, is actually a fractional part of that scene, just as the sensations and perceptions cited by Lenin are constituents of the objective reality surrounding him. This invalidates the opposition between inside and outside, subjective and objective, since each now becomes a constituent of the other. It is not that they are indistinguishable, no more than the sun and a plant are indistinguishable. But we now become aware of the fallacy of assuming that they are separate. Whether thoughts are consequences of reality (as the Materialists argue) or reality the consequence of thoughts (as the Idealists argue) each is part of the other and — to this extent — inseparable.

Our awareness of a thing is part of that thing

The continuity between things that affect each other must also exist between consciousness and all those things we are conscious of. We can only become aware of things around us as they impinge on our being in some way, when our sensory apparatus responds to their stimuli. Just as the sun and the plant become part of each other, so the sun and our thoughts about it are part of each other (since those thoughts are consequent on the existence of the sun and, therefore, part of it). It is this reasoning that allows us to claim our awareness extends to, and consists of, those things of which we are aware.

Intelligence in representation

In cinema theory, the mind of the viewer is doubled in the apparatus providing narcissistic pleasure — the screen. As with the mirror, when we ascribe beliefs to an object apparently outside ourselves then we acknowledge a kind of remote intelligence. In cinema the viewer ascribes beliefs to figures on the screen (which are essentially moving shadows) through a process of stereotyping, imagining and self-reflecting. It is now generally accepted that the same perceptual resources are employed in perceiving and imagining and, moreover, each perceptual mode competes for these resources. In effect it is more difficult to imagine at the same time one is perceiving and vice versa. It is also possible to confuse the two effects and think that one perceives when, in fact, one imagines. This condition might lend support to the idea that that mental images and representations are continuous with objects in the world, or at least that they overlap. It certainly confirms a quasi-truth about the act of representation that has it origins in the confusion between imagination and perception. In the process of representation, truth can rapidly become delegated to fantasy since the competition for cognitive resources is weighted in favour of the imagination by the (regressive) conditions of viewing. Character centred motivation, whether in the pointing finger of a Quattrocento angel or the behaviour of a film character, is recognisable from our real experience of common-sense stereotypes. Once the characters upon the screen begin to behave as though they are intelligent (planning and deferring action until it serves a goal) as distinct from reacting to stimuli (slapstick and spectacle) then the spectator's own intelligence is reflected back in a flattering representational form.

My awareness extends to, and consists in, those things of which I am aware.

Non-separate humans

Non-separate humans

To what extent do cultural experiences (such as movie-going where the mind and technology are so exquisitely convolved) reiterate the organic indeterminacy of the body's limits? On the face of it nothing would appear to be plainer than the existence of the separate bodies we inhabit or the unique minds we express, but this assumption cannot be sustained for long.

The human membrane

The human body, in common with most life forms, can be seen as a conglomeration of cells that is given a specific scientific status. In this epistemology they are organised into types according to function: nervous, muscular, skeletal, etc. The most visible are those that make up the *cutaneous membrane*, or the skin. As the name suggests, this amalgam of living, dying and dead cells forms a membrane around the organs and vessels, and performs a number of vital functions. It protects the internal tissues of the body from heat, shock, bacteria, viruses, water, dehydration and so on. It also acts as a principal sensory organ since it is richly endowed with nerve receptors for heat, cold, texture and pressure. The skin helps the body to produce vitamin D through exposure to ultraviolet light and aids regulation of the body temperature by adjusting blood flow and glandular excretion. The skin is also the site of a number of vital cultural functions beyond

37

the anatomical, most obviously in the way it defines race, religion, lifestyle and even social class. In many cultures, where great investment is made in its appearance, skin is the focus of adornment and display and part of the complex patterns of social signification. What these anatomical and sociological functions of the skin have in common is that they show it to be the thing that both separates us from the world and connects us to it. In the case of an individual cell the membrane functions to contain the cell contents and act as a permeable conduit between the contents and their surroundings. Whilst the human skin is literally a membrane that regulates our existence in the world it is also a metaphorical membrane that has an integral social signification (see note 2). All the functions of skin are concerned with how the human being negotiates reality (including other humans) and, more specifically, how they are integrated into it.

Models as complex as what they model

One immediate outcome of the ideas proposed so far is the redundancy many other, well-established, models of complex reality. In the postdigital membrane, the inter-reaction between culture and technology produces a floating amalgam in which self image, artefact or system and generalised representations of desire coexist — albeit in competition. This gives rise to inevitable tensions as competing forces vie for attention, just as competing desires vie to be satisfied. One outcome that is apparent in many historical accounts is a distorted perception of how technology and individuals co-operate since the inherent complexity obscures our view. Partisan accounts either vilify technology or denigrate the cultural impact of participation. An inevitable casualty in both camps is desire which is at best recast as a bio-erotic impulse exploited to shift the goods from the supermarket, or to draw the consumer through the door (or at least to the keyhole). This overlooks the fundamental influence that human desire has in shaping technology, which is greater in significance than that of the engineer, although this only becomes clear when one accepts the intelligence outside our heads. There is an equal danger of taking at face value technology's own description of itself. The cutting edge metaphors forged in the heat of technological discourses are often in contradiction with one's experience, and the fuzziness and confusion of desire. Attempts to over-ride the discrepancy between idealised technology and the disappointment of its inevitable shortcomings can diminish the breadth of the human experience and lead to a shallow, one-sided account. Such reductive descriptions regard technologies as distinct from humans at the same time as they dismember the head from the torso. As a consequence of releasing the mind and intelligence from the constraints of a specific site in the body, the human is revealed as a membranous tissue in which the unequal demands of desire, imagination and technology are stabilised, albeit temporarily. The advantage of the membrane metaphor is that it can embrace the complexity of the various determinants that shape perceived realities whilst avoiding the ossification and ahistoricism of conventional models.

The cinematic membrane

This section has tried to establish continuity between ideas, the active human body and aspects of the environment in which we exist. If we acknowledge this continuity it allows us to recognise and describe instances of encoded consciousness outside of ourselves; facets of thought deposited in non-organic substrates such as those embodied on walls or the cinematic apparatus. Such mechanisation of our mental experience is particularly visible in the technology of Classical Hollywood Cinema which, like a mirror, reacts to the environment it partially creates — absorbing and reflecting the things we imagine and desire. The cinematic apparatus may also be thought of as a membrane — an indefinite space that both separates and connects (fragments and unifies) the experiences we perceive or imagine. In the case of cinema, one experiences a contrast between actual, lived existence and the compressed, hyper-reality of the mise-en-scène. It is this contrast that the cinematic membrane both asserts and embraces.

Logo from admission ticket for Tintern Abbey, UK

Section Three

PROPOSITION:
Histories, like predictions and superstitions, are narratives of human imagination

What aspect of realism justifies humans with wings?

Spatial amputation

As we know, the artists of the Quattrocento had various methods for representing convincing depictions of space including the convergence of parallel lines, overlapping figures and various geometric laws that were applied piecemeal to the picture. The significance of Brunelleschi, perhaps the most well-known, was that he devised a system that not only could apply to any spatial regression but also was founded upon verifiable geometric calculations which were independent of the artist. His demonstration of the system is well-documented and comprised paintings of two buildings in Florence, the Baptistery of St. John and the Palazzo de' Signori. To heighten the effect of his painting of the Baptistery, Brunelleschi defined a precise subject position by drilling a small aperture in the panel and obliged the viewer to view the image in a mirror spied through this hole from the back. By this arrangement the viewer, reduced to a single eye, was placed in more or less the same relation to the image of the building as the fully sensible artist had been when he produced the painting. The spectacle and lure of the image, its illusion of reality, ameliorated the amputation inflicted upon the subject by this unequal relation. For the second demonstration he cut away the sky from the painting so that the image could be held between the viewer and the Palazzo de' Signori so that its representation could be verified. In this picture too, the 'real world' was reduced and partially obliterated by its representation.

The vanishing point – the reduction of space-time

There were of course many alternative ways of depicting the environment as a space in which events happened. Since 1300, at least in the West, the dominant convention used to depict the world truncated the observer's perception of reality (all images were silent, monocular and static). This vision of space, in which infinity was conceived as a single point, accorded with the prevailing representation of history. Here, non-present time was posited as a consistent regression, both backwards and forwards, in which events were ordered causally. Such causal narratives demanded that the present be truncated since it could not be appreciated for its immediate complexity but only as the destination of the past and the starting point of the future.

The history machine; temporal looping

Terry Gilliam's film, *Twelve Monkeys* is an over-elaborate fantasy of a technological dystopia set forty years on from the present. Towards the end it contains fascinating twist when James Cole (Bruce Willis), who has been sent back and forth in time, makes a telephone call from 1996 to scientists in 2035 — his year of departure. It is a moment that is supposed to rectify a cardinal error in efforts to research the history of the virus which devastated mankind and report back to the diminished civilisation that remains. Even with his low-IQ he realises that what you see depends on what you are looking for, and if you set out looking for the wrong causes of an event, you will find the wrong explanation. As Cole rings off he finds the recipient of the call to the future standing beside him. Apparently the technology for time travel has become so sophisticated in his brief absence from 2035 that projection into the past time and space is possible with pinpoint precision. James Cole, the time-travelling convict whose misadventures constitute the greater part of the film, had become so accustomed to the laborious and haphazard process of travelling back in time, missing the point and then lurching forward towards an approximation of his destination, that he is understandably astonished by this exponential leap in time-travel technology. His antagonists, it seems, can find the exact second and precise location of a happening in the past and send a representative there to observe it. It may seem like the historians dream-machine, but, as the film makes clear, a fumbling grasp of events is sometimes more likely to yield an understanding of the broad range of historical determinants than the laser-sharp recall of the clinical methodology of 2035. The story-line of the *Twelve Monkeys* plot celebrates the productive inefficiency of the ordinary man over the unproductive precision of the scientists. It also serves as an allegory for the virtue of the film-strip and videotape, despite all that rewinding, frame counting and laborious edge numbering which has apparently been shown the door by digital media. It is a cinematic caution that non-linear editing means 'rewind' and 'fast forward' are indistinguishable from the 'play' mode, and also that the flashback can no longer unambiguously mean 'I remember'. Quite typically for a mainstream film, *Twelve Monkeys* shows the digital age of the near future as a time of material dereliction, totalitarian political regimes and moral deadlock. As with many recent films, advanced technologies are cloaked in neo-fascist decor and presented as terribly precise, digital rather than analogical, inhuman and above all wrong-headed and certain to miss the point of history.

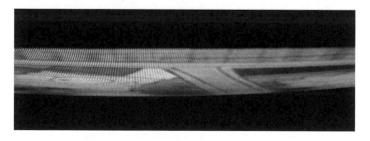

Homer and Bart driving West, vanishing

We can't remember the present, or predict it, but we can remember the future and predict the past.

In common with many sci-fi fantasies *Twelve Monkeys* affirms that the only substance the past and the future have is their manifestation in the present where each remains a set of uncertain probabilities. It is inadequate, however, to imagine the present as being singular, instantaneous or lacking in dimensions — a kind of temporal vanishing point. The present is not only a compound of memories and anticipations (we oscillate between projecting backwards and forwards) but also a continuous moment without hiatus that has depth as well as height. In other words, the present is a space as well as a time. It is within this space-time that we manipulate our memories of the past and future, construct our narratives and create our recollections. Yet these narratives can only incompletely represent what in actuality happened or will happen. The compulsion to fantasise, to extrapolate, to re-order, to impose structure upon the chaotic fabric of experience seems so universal that it appears to be almost a requirement of civilisation. Yet our recollections and predictions have dubious reliability since they can, of necessity, only be partial and reductive. This is not to say that memories of real events are not, in themselves, real. Nor is it to say that our expectations of the future are never vindicated. Rather, it is to recognise the contingent nature of prophecies about the past and the future. History and futurology not only share narrative tactics but also converge and convolve in the same 'thick' space of the Now.

Shaping the present

It is now generally accepted that just because two things can be distinct in language it does not follow that there is a distinction in reality. Space and time, at least in practical scientific discourse, are regarded as inseparable. As a consequence, we can think of the present as a 'space-time' that expands or contracts according to the density of events and in which the past and the future are manifest but never resolved. Using this conceptual tool to manage the complexities of reality we are relieved of the need to determine how the past affects the present or how the future is derived from the past. We now have a continuous space in which all tenses are concurrent, and out of which our histories and predictions emerge. Everything is expressed (and occurs) in the space-time of the present. This multi-dimensional conception of the present can be visualised thus:

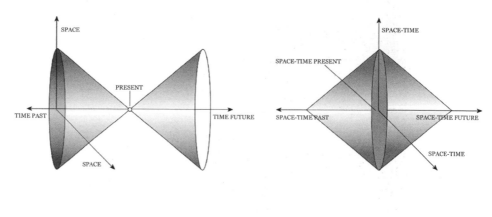

Figure 1   Figure 2

Two ways of visualising the present. In Figure 1 (derived from Stephen Hawking, *A Brief History of Time*) the Present is visualised as a singularity that is the inevitable consequence of Time Past. Time Future, therefore, is the inevitable consequence of the Present and the Present is the collapse of all possibilities. In Figure 2, by contrast, the Present is the expansion of all possibilities and exists as the eternal Space-time that mediates our conception of Past and Future events — a conception that always occurs in the Present. The possibilities of the Past and the Future collapse as we try to order them from the position of the Present. This could be characterised as the postdigital conception of time-space.

The present membrane

In the postdigital conception of time-space (figure 2) our present becomes a spatiotemporal membrane, an involuntarily condition in which we witness the bi-directional flow of events from one side of the temporal axis to the other — with access to neither side. This demands nothing less than the recovery of those perceptual possibilities lost to the 'realism' of linear perspective and linear narratives.

44

Image without a vanishing point

The present is eternal

The competing cultural and economic imperatives that compel us to desire the future or recover the past tend to reinforce the universal and eternal present as incomplete and temporary. We always imagine ourselves to be going forward (advancing/anticipating) while actually aching to return (nostalgia/retreat) and expecting everything to change (although, as we have pointed out earlier, things remain largely the same). At times we seem reluctant to accept the fact that we are eternally locked in the present — we fail to see that all conceptions of past and future events lie within the immediate moment. The present becomes something that we must resolve at another time or imagine away. To aid us in this disavowal of what is 'now' we develop ways of freezing, recapturing or postponing the present that are manifest as new inventions — technologies of distraction or improvement. For example, the widespread compulsion to capture the 'now' so that it can be experienced at will in the 'future' has (at least in part) provoked inventions like painting, photography, cinema, sound recording and video. Such technologies are also used to recapture the past (retrophilia, heritage industries) and construct a

45

vision of the future (sci-fi, futurology, propaganda) to try and escape the present. This consumption of recorded material from the past co-exists without contradiction with the desire for immersion in a technological future — a double disavowal of the present.

Grey rainbow

Degraded projections

The inability of time capturing technologies to completely escape the gravity of the present become especially clear in attempts to simulate the past or the future. American futuristic sci-fi films of the 1950s inevitably betray the behaviours, aesthetics and preoccupations their period, like a bad wig betrays baldness. Similarly, the authenticity of British historical dramas of the 1940s is undermined by the cosmetic priorities of the era (bow lips and ringlets for the women, the mannered postures of men in Gainsborough Films). Just as futuristic visions from the past are inevitably inflected with their present, so our present readings of past artefacts can never overlook the unintentional layers of meaning that accrete upon them as they age. The degradation, low resolution, scratches, dust, compressed audio, jump cuts and head-wear of moving and still images all insist on the frailty of the medium and remind us that we are receiving them in our own time. No media are impervious to such mutilations, even those chosen for their resistance to temporal ravages must eventually succumb through misuse and weathering (a process that often enriches, rather than diminishes, their meaning).

False memories

Unsurprisingly, a sub-industry has evolved in movies and broadcasting dedicated to simulating the natural ageing process of film and video to meet the ever more sophisticated demands of producers and audiences who can tell the difference between Super-8 and Hi-8. This strategy seems symptomatic of a desire to recreate the past, not as it actually was but as it is falsely perceived through the media that captured it (hence the tendency to think that all pre-1970 events occurred in monochrome). Benny Hill merged a stilted perception of early silent cinema (burlesque, accelerated projection) with the technical possibilities of 1970s television by using speeded up VT, thereby overlaying overt erotic desire with an air of lost innocence. It is inevitable that the hi-tech recording apparatus of today will be similarly parodied in times to come.

The past in the present

What keeps us going?

Attempts to re-capture the past or bring forward the future play with our imaginative resources and, as we have said, divert our experience of the present. We undoubtedly seek to reconcile the conflicts between what we experience in the present and what we imagine — between realities and desires. Experience consists in what actually happens to us whilst imagination contains our hopes, wants, fantasies and dreams. Our expectation of resolving the discord between the two is based on the assumption that there is some projected time at which conflict is resolved (both in retrospect and in prospect). It is this assumption that many political slogans capitalise on ('Forward to the Future', 'Building a Better Britain', 'Back to Basics' etc.). In spite of the high failure rate, we are encouraged seek solutions to our problems because of the occasional successes in resolving conflicts and ameliorating threats — the war ends, a cure for smallpox is found, we can treat the symptoms of Aids. What keeps us going is the hope that present conflicts are resolved in the past or future.

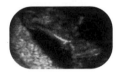

The narrative vanishing point

Canonic cinematic narration exploits this generalised desire for resolution *ad infinitum* as well as reinforcing the idea that (with the right moral stance and facial features) all ruptures can be resolved — at least sufficiently to allow us to leave the movie-house. It provides the basis for the structural conformity of Classic Hollywood Cinema that enables similar and coherent readings of the same material by internationally dispersed audiences with different prototype schemata. There are a number of distinct features of canonical narration, most notably: character centred causality, goal-orientated action, and double plot line. Closure is achieved by the resolution of these causal structures either sequentially or simultaneously, with one usually involving heterosexual couple formation whilst the other involves the initial cause and/or the revelation of truth. The overarching tension in the plot reaches conclusion when the deadline expires and goals are either achieved or lost.

The awkward body

Plot resolution satisfies the human desire for reconciliation by utilising a common-sense construction of events based on the illusions of moral symmetry and causality. This much has been well-established by orthodox cinema theorists who describe the cinematic experience largely in mentalist terms. More radical, however, is the demand for a model of history rooted not in socially constructed ideologies but within the amputated sensorium we know as the human body — it is the body as much as the mind that demands resolution and the healing of ruptures. By reinserting the body into the structure of experience we are rectifying a cardinal omission of psycho-linguistic theory. In postdigital terms, the mind must necessarily include the body with its responses, emotional spasms and varying states of vibrational intensity. Cinema, of course, demands as much physical engagement as it does psychic engagement, despite the apparently disembodied experience of viewing.

Theories of collapsing reality: The apparatus of mind and machine

Just as representation in the cinema and in certain paintings reduces human experience, so attempts to recover it are prone to favour one dimension of perception at the expense of another. The difficulty may lie in the very structure of the brain, or at any rate, our mental apparatus. There appears to be sufficient evidence that this apparatus operates according to two distinct paradigms, the logical and the gestalt. The detail need not concern us except in two general principles. Firstly, intelligent behaviour is the product of the logical in which a causal chain sequentially links evidence to produce a conclusion and the gestalt in which all information is examined at the same moment. The second principle is that there appears to be specialised apparatus for each of these two modes of knowing that are integral to our mental operation. The all-perceiving subject is a gestalt model that then appeals to psychoanalytic theory to effect a bond with the logical. Whether or not these principles provide a secure foundation for human psychology, they do seem to support the hypothesis presented here, that, given a contradictory picture, humans tend to privilege overall coherence and logical simplicity over the complexity of lived experience (see note 3).

49

What do we want?

It is nothing extraordinary to point out the human desire for a resolution of the conflict between desire and reality, or the collapse of confusing complexity into a comforting state of order — many thinkers have described it more eloquently than has been done here. Nevertheless, it might be the case that we have underestimated its importance in the general understanding of human nature. Or at least this may be so if we include within our understanding of 'conflict' states such as longing, pain, fear, loneliness, sadness and anxiety as well as bodily homeostatic imperatives such as cold, thirst, hunger and heat. Through living in the world we are moved to action and thought by things that require attention or elicit a response. How do we then influence things in our favour, how do we make the world accord with our desire or need? It is likely, in many cases, that we mentally imagine a resolution prior to achieving one in reality. The imagined resolution would then guide our actions in bringing forward the future (going to the fridge to get a cold drink). This ability to think forward and imagine solutions, which may be unique to humans, would certainly contribute to our evolutionary success. It would mean that our needs and desires were met more effectively — that we were more able to turn events in our favour, although our cats, by and large, seem to live better lives than those of us who feed them.

In(tegrated)-determinants

The conflicts between desire and reality so far described could be termed 'psychological' or 'biological' and would count among the many drives or instincts that motivate human behaviour. Rather than trying to unpack all these here we are pointing to a general interaction between desire and imagination with its consequent effect on reality — especially the reality of history, or at least our perception of history. The most useful model of historical change has much in common with one of the least scientific of apparatuses — the 'Ouija Board'. As the *planchette* or glass moves between the letters, no single determining force can be identified. Its movement is the product of a network of competing desires and pressures from the participants as well as the undeniable forces of gravity, humidity and the resistance between the surface of the board and the planchette. Isolating one of the vectors will not only fail to explain the ultimate sentence that is spelled out, but will inevitably lead to a misleading aetiology.

Isolating one of the vectors will not only fail to explain the ultimate sentence that is spelled out, but will inevitably lead to a misleading aetiology

Reality recedes through perspective

The temporal and spatial present is a site of loss and amputation. 'History' and 'Prediction' are institutional narrative forms that postpone the present in favour of a mediated representation where possibilities have collapsed and uncertainties are resolved. In the same way, the continuum of the visual spatial world in flux is rendered as a truncated illusion by pictorial representation. The consequently enormous reduction in the perceiver's self image can reinforce a sense of isolation and fragmentation that renders the present incomplete and unsatisfactory — hence the tendency to defer the now in favour of an imagined past or the future (using an array of technologies). These complex and contradictory movements constitute what we understand as the permeable membrane of the present.

A membranous history — Where Was I?

'In November 1963 I was standing on a railway station in London returning home after some heavy adolescent drinking when I heard that John F Kennedy had been shot. I bought a newspaper, which I think I left on the train. Later, when I was grown up, I played the Abe Zapruder footage of the shooting over and over again, tormenting my students with questions about media specificity, I was always unable to reconcile my experience with what I saw on the screen. The horror of Jackie Kennedy crawling across the trunk of the car away from the blood had no bearing on my personal memory of the event.

Early one morning 30 or so years later, I was standing barefoot on quarry tiles in the kitchen, waiting for a kettle to boil when I knew that something was wrong. Instead of the usual baiting of politicians and the carping criticisms of European Union, the radio was telling me about the Royal family, Princess Diana and Dodi Fayed. Then 'for those of you who have just tuned in' (a phrase to be repeated at ten minute intervals) the slim facts of the news of the crash in the Pont d'Alma tunnel were reiterated followed by more family detail. In the intervening years the world had become electronic and instead of buying a paper I switched on all my radios and televisions just to hear it all over again simultaneously. Perhaps because the crash was in a tunnel out of sight, certainly because all stations were using a single news source, broadcast media suddenly seemed inadequate; somewhere above it all in a kernel of ethereal technology which fused image sound and text one message beamed up from central Paris. There was no opinion, no post-modern collapse of history; the one story had to be reread like a love letter or legal summons as though a concealed truth was embedded in the punctuation.

I was indeed back on the railway station, a slightly fuzzy, reluctantly passive consumer of the news making up my own story. None of this reaction was especially unique to me nor evidently were my subsequent responses of (i) finding someone to tell who did not know (ii) telephoning someone else to see if they knew (iii) logging onto the Internet to check Reuters and CNN in order to see what I was not being told and regain some control of my own opinion. These news agencies of course were calling the shots and quickly it was clear that a web search was called for to see if there was any inside story, which the lawyers had repressed. The first fifty hits for 'Diana Princess of Wales' that the search brought up were 'fan' sites sprinkled with obscene and erotic content — this was shocking to me although, as a web journalist perhaps it should not have been, but it simply had not occurred to me that anyone would be interested in her in that way. Within hours however these sites quietly dropped off the search engines and new sites opened in which people expressed condolences, shock and sadness. Later new pages began to appear which contained conspiracy theories ranging from the simple to the elaborate and the plausible to the apparently outlandish (e.g. establishment cover up of Diana's abduction by aliens). It appeared that the more unrestrained reactions to violent death and all the more dramatic ideas which the broadcast media were constrained by law and convention to leave unspoken found expression in the less regulated public space of the internet.

As the week progressed more extraordinary things happened most of which were so well represented in the press that they do not require extensive exposition here. Suffice to say that huge crowds gathered outside Kensington Palace, and many new arrivals brought a gift of flowers with a message or personal mementoes — soft toys, religious images, candles etc. — to leave at the gates. Visitors tied cards and favourite photographs to the railings, queued to sign the official book of condolence and whiled the day away distracting themselves reading the personal thoughts of others posted in a moment of apparently private reaction to the sudden news. The Palace opened an internet website and was obliged to break further with tradition by lowering the flag on Buckingham Palace in what was interpreted as a response to people power. Initially the crowd comprised mainly a socially defined group of middlebrow Royalists, and a working class at home with the idea of national identity, and lonely people at home with the myths constructed by the Princess's public relations office. Republicans, intellectuals, the 'chattering classes' and the more restrained sectors of British society stayed away. All the past and the future of 'Diana the soap opera' was collapsed in a moment on 1st September as causes for the accident failed to materialise. It was a shock, and gradually those who in the normal course of events saw the Diana cult and her subsequent media hagiography as yet another expression of manipulative bourgeois ideology, were drawn to the fascination of the crowd and began to subscribe to the myth. Their natural immunity to the schmaltz and teddy bears dissolved in favour of the reduced sensibility demanded as the entrance fee to the national community who were coming to terms with an event without explanation.'

Coming to terms with an event without explanation

Section Four

PROPOSITION:
Language is a technology through which human desire is satisfied and generated.

Desire insists on the presence of the absent

Desire is that which is manifested in the interval that demands hollows within itself, in as much as the subject, in articulating the signifying chain, brings to light the want-to-be, together with the appeal to receive the complement from the Other, if the Other, the locus of speech, is also the locus of this want, or lack.

Jacques Lacan, *Direction of treatment and principles of its power*

Desire in/out of Language

For Jacques Lacan, and many of those who followed him, 'the Other' is part of a signifying chain that exists within Language. Indeed, he is one of the main proponents of the view that consciousness, the unconscious and desire are 'structured as a language.' Here, Lacan represents the school of thinking that emerged from the coalition of Linguistics and Psychoanalysis and dominated Western intellectual activity for most of the latter half of the twentieth century. For many of these thinkers there is nothing 'outside' language, or at least nothing one can talk about, since all meaningful mental activity is linguistic. One would not want to deny the primordial importance of language in all aspects of human life. Indeed, those labelled 'Structuralists' and 'Post-Structuralists' have a point in that it is almost impossible to describe any aspect of human existence without recourse to some form of words or signs. However, should we then take it that all consciousness is language based, in the sense of verbal, symbolic language? Or should we leave open the possibility of 'non-linguistic' or 'extra-linguistic' awareness or thought? The question is a vital one. Does an impulse like 'desire', for example, arise exclusively from within the structuring protocol of language (as Lacan might suggest)? Or, alternatively, is desire normally subject to the social regulation of language but essentially prior to it — something greater than we words we can use to describe it? To extend the point, would we still have desire if we didn't have language?

Gaining language, losing love

According to some descriptions of language acquisition there is a clear distinction between the state of being in a 'pre-linguistic' child and one that has acquired it. To many psychologists, the pre-linguistic child main-tains a state of 'ideal unity' in which the world and themselves form and unbroken continuity. At some time during or after the pivotal 'mirror-phase' this unity is disrupted by the realisation that the self is distinct from the world and the ego emerges. This 'Splitting of Subject' is fundamental to the acquisition of language since language operates on the basis of differences — the distinctions between things that we make when we name them. The resulting condition is one of both loss and gain. Lost is the holistic bliss of imaginary identification with the mother, but gained is the agency of language, the power to identify objects and symbolically communicate. For most psychoanalysts and psychologists this is a crucial moment in development — a point of no-return that sets up the conditions for all further linguistic and social integration. The significance, for our purposes, is the process of differentiation that unfolds as one's language evolves. It is through this that the illusion of a world made up of separated things is reinforced, although the things 'in themselves' are actually continuous. For example, we differentiate trees from the ground in order to name each of them, creating the illusion of two things where there are/is (n)one. Whilst our sensory apparatus is able to distinguish between shapes, densities, textures and patterns it is our linguistic categorisations that encode these apparent differences into words. And being thus encoded they become structured as a series of opposing signifiers (to adopt the structuralist jargon). Consequently, our conscious experience of the world is fragmentary.

Becoming whole (again)

Some of the ideas about reality and the self that emerged around the beginning of the twentieth century reflected a general revision in the way that the world was seen. What emerged was a more fragmented and varied material and intellectual environment than had existed before. It was a period that saw, amongst other things, the introduction of telephones, cars, aeroplanes, gramophones, cinema, electrical power, machine-guns and x-rays. In this period rich in new objects and ideas there were marked conceptual shifts in contemporary physics, literature and philosophical speculation about the mind. Various disparate movements in painting and sculpture synthesised and gave expression to these changes. Cubism, Futurism, Vorticism, Rayonism, and Dadaism were all stylistically recognisable affirmations of quite new and distinct beliefs about the self and the world. What they shared, however, was the idea that the fixed object had become the ambient object moving at speed and the Self, in relation to the material, could no longer be held as absolutely distinct, becoming instead synthetic and provisional. One only needs to look at Cubist figure-paintings to confirm this. In one more instance of desire shaping technology, Modernism posited a fractured view of the self and the world that reified the concept of alienation, the essential de-coupling of oneself from one's surroundings (a threat represented elsewhere as the 'March of the Machines'). These artistic movements resurrected a much older, deep-seated malaise of the human condition (the divided-self) that now benefitted from scientific recognition, and which became rationalised through the emerging Psychoanalytic method and various revolutionary ideologies.

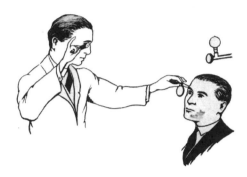

Explanations of human desires

Explanations of human desires

Theories such as Psychoanalysis, Structuralism, and Modernism, which emerged from the intellectual turbulence of the last century, inevitably inflect much of what can be said today. One must take account of them if only because they have gained a certain irresistible authority over the discourse of desire, representation and language. Yet for all our Modernity we find ourselves saying what has been said before. Lacan, the psycholinguistic theorist, is describing above (in somewhat technical language) something that has apparently been understood about desire for many millennia. Sigmund Freud can find no better explanation for the origin of the sexual instinct than to cite Plato. In the *Symposium*, Aristophanes says:

The sexes were not two as they are now, but originally three in number; there was man, woman, and the union of the two... the primeval man was round, his back and sides forming a circle; and he had four hands and four feet, one head with two faces, looking opposite ways, set on a round neck and precisely alike; also four ears, two privy members, and the remainder to correspond...the gods could not suffer their insolence to be unrestrained. At last, after a good deal of reflection, Zeus discovered a way... He spoke and cut men in two, like a sorb-apple which is halved for pickling... After the division the two parts of man, each desiring his other half, came together, and throwing their arms about one another, entwined in mutual embraces, longing to grow into one, they were on the point of dying from hunger and self-neglect, because they did not like to do anything apart; and when one of the halves died and the other survived, the survivor sought another mate, man or woman as we call them, — being the sections of entire men or women, — and clung to that.

Plato, *The Symposium, The Speech of Aristophenes*

Freud, of course, had deep respect for the powerful metaphors embedded in such myths and the insights they offer into the human condition. In a note that refers to the genealogy of this Platonic explanation of sexuality, Freud recognises that "essentially the same theory" appears in the *Upanishads* which predated Plato and might well have migrated West from Asia.

But he felt no delight. Therefore a man who is lonely feels no delight. He wished for a second. He was so large as man and wife together. He then made this his Self fall into two, and then arose husband of wife. Therefore Yagñavalkya said: "We two are thus (each of us) like half a shell." Therefore the void which was there, is filled by the wife.

Quoted in: Sigmund Freud. *On Metapsychology: The Theory of Psychoanalysis.*

In *L'Amour fou* the surrealist writer André Breton reiterates the union of desire and loss. He recalls a visit to a flea market with the sculptor Giacometti where they examined a number of objects including a (presumably African) metal mask and a phallic decorative spoon, which Breton bought. It later transpires that a lost lover of Breton's was in the same place and saw him examine the mask which she, earlier, had also shown interest in. Breton writes:

> Although she was intrigued by the mask, she had put it down as I had. "Of Eros and the struggle against Eros!" My disquiet, and perhaps hers before mine, in front of the mask — about whose use I later had such painful information — the strange figure (in the form of an X half dark, half bright) formed by this encounter I was unaware of but she was not, an encounter so precisely based upon such an object, led me to think that in this moment it becomes the precipitate of the "death instinct" dominating me for so long because of the loss of a beloved being, as opposed to the sexual instinct, which, a few steps farther on, was going to be satisfied in the discovery of the spoon. There could be no more concrete a verification of Freud's statement: "The two instincts, the sexual instinct and the death instinct, behave like preservation instincts, in the strictest sense of the word, because they tend, both of them, to re-establish a state which was troubled by the apparition of life." But I had to start loving again, not just to keep on living!

André Breton. *L'Amour fou.*

Love (or desire) is seen here as a drive toward unification, completion, wholeness which, however briefly achieved or tasted, returns us to some past state that we have lost — Preconsciousness? Self-annihilation? Death? — a sentiment rehearsed in *Lolita* as Humbert reconciles his own paedophilia through the death of his adolescent love, Annabel, when he was 13: "In point of fact there might have been no Lolita at all had I not loved, one summer, a certain initial girl-child." (Nabokov). Of course, we know that love and desire often find their strongest conventional expression in sexual contact, or even the repressed desire for sexual contact as Freud would argue in the case of children and parents (Oedipus, et al.). It hardly needs stating that the extraordinary diversity of ways in which humans express desire, from tender to depraved, from platonic to erotic, testifies to the power and depth of this drive. Yet in the myths and theories that have abounded and amongst all that has been said about it, a simple principle appears to remain constant in accounts of human desire — the will to unity, the will to regain the whole. Through sexual union they suggest that we come, perhaps, as close as ever to "receiving the complement from the Other" by, as far as possible, joining with the Other as one.

Non-linguistic desire

Whilst one might take issue with the autonomous view of language proposed by the Saussurian/Freudian paradigm, it does at the very least provide a provisional description. In such accounts, as we have said, language enforces a fragmentary view of the world that can only be negotiated through the use of language itself (the rules of which we have no choice but to follow). Desire, then, emerges in the struggle to reconstitute the wholeness that language has splintered. However, there are two deficiencies in this model: it does not allow for extra-linguistic or pre-linguistic thought or desire and it relegates the role of the body in the operation of mind. In this paradigm, thought and desire become reduced to mere symbolic processing of a kind that might neatly be replicated in digital computers. If it is the case that the operation of language is responsible for the manifestation of desire (as Lacan suggests above) this would not account for the pre-linguistic desire of small children. Nor would it account for the nagging demands of our bodies as a consequence of their integration with their surroundings.

Outside language, desire outside our minds

Whilst language may powerfully reinforce a fragmented world, and mediate our subsequent desire toward reunification, it cannot be held entirely responsible for all our conditions of desire. Nor can it always repair the damage it does. In the postdigital model, language (symbolic processing) is seen as only one of the many methods that humans have employed to remove obstacles to the satisfaction of desire, to regain continuity with or gain control over the world. Indeed, the very evolution of language may be a function of the advantage gained by the ability to express human will. Insofar as language is seen as the primary vehicle of fragmentation, the imperative of many technologies has been to loosen the grip of language over thought. We know that techniques have developed (meditation, drugs and ritual to name a few) that attempt to bypass, or suppress, the common operation of linguistic thought so as to negate the fragmentation that language insists upon. Meditation, for example, is the concerted attempt to stop 'thinking' altogether. Some rituals use chanting where repeated use makes words or phrases 'meaningless.' Certain narcotics can induce states of wellbeing entirely independently of any linguistic activity. In each case, it is not the mind alone being transformed. Mental and physical states of being are inextricably linked and sensations can only occur in the body (the brain being devoid of sensory nerves). Therefore, desire is not confined to a mental symbolic lack but also includes the unequal pressures exerted around the membrane of the body, where the body is recognised as conscious, subject to and implicated in the activity of the environment. As a result, desire, like intelligence and consciousness, must extend beyond the immediate confines of the brain to include the body, its actions and those 'external' objects we invest desire in.

Language as technology

If we regard language as continuous with other technologies then, to some degree, it must operate in comparable ways. Perhaps looking at similar technologies may help us to understand more clearly how language functions relative to imagination and desire. Postdigitalism proposes that language be understood as a means to an end — the on-going attempt to satisfy a desire that precedes language, but which is also generated by it. Operating within the sphere of imagination (the ability to see what is not there) language is a tool that increases the probability that desires will be met.

Both perception and imagination contribute to our sense of reality

Extended continuities

Regarding language as continuous with other technology accords with orthodox methodologies that search for continuities between ideas, or things that were formally seen as separate, such as humans and machines. Bruce Mazlish has argued that a conceptual shift is necessary for humans to see thenselves as indistinguishable from technology and that this may follow on from other shifts that have occurred in recent history. The continuity of the Earth with the rest of the Universe that Copernicus proposed was followed by a second continuity arising from Darwin's conclusion that man was biologically continuous with other life forms. The third continuity, as Mazlish sees it, was Freudian psychology's suggestion that the indigenous European adult was continuous with the child and other cultures. It may be that, in the postdigital era, we are ready to accept a deep continuity between all things that appear to be separate. In particular, the continuity between human and machine will no longer seem fantastic and, indeed, can be regarded as being well established. An example of the fusion of humans and machines we have already discussed is the operation of the cinema, which induces a deep psychological intimacy between the viewer's mind and the chemo-mechanical apparatus. In classical narrative cinema this intimacy creates a sense of coherence and unity that is largely absent from mundane life. The purpose of cinema, as a technology, is to satisfy desire (whilst also generating it) and, in this way, it mirrors the purpose of language.

Why do we watch movies?

One might regard the Classical cinema, with its artifices of Technicolor, Cinemascope and stereo sound, as a machine the purpose of which is to fully command our senses whilst excluding all ecto-cinematic sensation. If successful it forms in the imagination a finite and unified consciousness over which we have no self-reflexive control. It is only when the machine is poorly made — transparent or hackneyed plots, clichéd dialogue, stiff acting, shaky sets — that the grip over our consciousness is relaxed and we are able to gain some critical distance with which to catch our imagining selves at play. At its best however, cinema (in common with some other representational forms like theatre) requires total continuity between consciousness, body and machine. Without the direct emotional feedback provided by the enzymes and hormones of the living torso (fear, arousal, humour, joy) the unified consciousness would disperse in a sea of minor distractions. In its most convincing mode the cinematic apparatus utilises a utopian eye to represent its own vision and shows only that which can be seen by it, either all at once or by the progressive accretion of knowledge. It processes information into depositories of stored consciousness that the viewer can appear to manipulate with their own mind but with which, in fact, they become fused. The economic independence of the viewer's mind is entirely subsidised by the on-screen procession of information-rich data. It seems that cinema as a machine creates an idealised cinematic body, a coherent subject in which the mental and material constituents that (it is assumed) comprise the human being are temporarily unified in state of grace. The cinema provides a perfect example of non-linguistic desire.

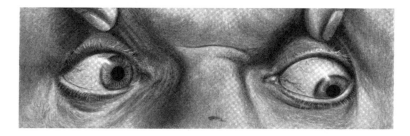

Non-linguistic desire

Edited perception

Contemporary theories about human perception lend support to the idea that we habitually constitute a holistic image from partial, even contradictory, data. If imagining and perceiving are separate functions that may get confused, then this confusion may be permuted differently between our five sensors simultaneously. The most privileged sensor, the eye, is prone to betray us. It is merely an extension of the visual cortex of the brain that filters to provide highly selective information from the available data. Our visual perception of the environment, no matter how real, is only a small portion of that reality. Our restricted point of view, our experience of twilight vision, sun glare and snow blindness remind us of the

64

limitations of our visual apparatus. Examination of visual perception in animals suggests that eyes are goal-orientated receptors that detect different portions of the electromagnetic spectrum relative to specific needs. The implication of this schema is that the senses act to fragment a holistic reality that is then partially reconstituted by the mind as a (mostly) unitary experience.

The act of splicing cuts and joins in order to reconstitute that which the camera has fragmented

The apparatus of film-making, like language, slices up reality and then reconstitutes it as a coherent whole according to its own rules, which we must subscribe to if we are to understand the film. However, as well as satisfying this desire for a coherent view of reality, the movies also generate specific new desires which only they can satisfy. During Hollywood's heyday of the mid-twentieth century two kinds of motorised editing tables were used. Both allowed the editor to synchronise precisely sound and image for lip-sync. This was a technical improvement on earlier hand operated systems which relied on the operator moving the film by hand across a light box (or later winding reels) at the correct speed, matching numbers on the film counter with those on the tape. One unforeseen consequence of this improvement was that the shots and sequences could be organised more rhythmically since what the editor saw was precisely what was projected for the audience. The more robust of the editing tables, the Moviola, was invented in the early 1920s by a Dutch/American engineer, Iwan Serrurier. In its later form it was a vertical machine operated by heavy duty foot pedals and fist-sized buttons. It made an alarming rattle as it ran the film, and a satisfying thump as it stopped at the right frame. Partly because it often damaged the film a smoother and more tranquil solution was found in European horizontal editing machines similar to the now familiar Steenbeck, on which the film swishes across the table and somehow stops at the right place without the spine juddering clunk of the Moviola. Both were used in Hollywood in the post-war period, and it is one of the enduring myths that Westerns were edited on Moviolas, whilst the melodramas and romances were cut on the smoochier flat beds. It conjures a view of the editing suite with, in one corner, macho men in check shirts and rolled up sleeves slamming at the controls to make sure that although the bad guy always drew first he somehow stopped a slug in the heart; and in the other, languorous men and women draping themselves over flat-beds, gently caressing levers and sobbing quietly for loves that might have been. This may be no more than a Hollywood folk tale, but it conjures up an image of the movie business that is too seductive to be troubled by mere truth. What is more, it has the compelling ring of commonsense about it since we what we say is invariably inflected by what we use to say it with. It suggests that technologies used to represent the world as 'joined-up' are anything but neutral, and the same linguistic cuts processed through different machines can mean very different things. More important, however, is that the very act of cutting seamless reality into units is necessary before we can represent it as a logical whole. Since this is an illusory and incomplete 'whole' (being reconstituted from the position of an amputated viewer) it retains multiple invisible elisions which generate new desires.

Representations invoke the presence of absent things, thus meeting and creating desires

The cinematic whole

Cinema offers an escapist alternative to the mundane chaos of everyday existence through spectacular dangers, adventures, personalities, shocks or intrigues that few of us would willingly expose ourselves to in real life. It offers on demand (indeed, we are promised as part of our movie-going contract) a condition that we seem to desire — our own coherent, unified consciousness, albeit one that is only on loan from the production. Each unique, incompatible and contradictory impulse in our being is wilfully suspended in a consensual, hallucinogenic soup that, like burgers, tastes pretty much the same everywhere. In 'apparatus theory' the success of a movie depends not so much on the fact that we watch it for pleasure, or even that we are engrossed in it, but to the extent that we become continuous with the apparatus. For most purposes we are indistinguishable from the movie as it plays and there is little spare to claim as our own. When the impulses and reactions of our bodies are drawn into this contract with the machine it becomes clear that the cinema is but an extended human organ.

The cinematic subject

Serious attempts have been made to analyse the 'reality effect' of cinema and to account for it by application of scientific knowledge, particularly that derived through the Freudian project. As we have said, recent work in the slightly 'harder' cognitive sciences suggests that the same perceptual resources are employed in imagining the world and in perceiving it and, for this reason, it is possible to confuse the two experiences. The importance of this discovery is that lens-based representations do not necessarily resemble 'live' perception but confirms only that the same nodes are triggered in the perceptual apparatus as in the act of perceiving. The cinema screen supports an illusion in which the absent is also present, where imagination becomes almost identical to perception. Moreover, the screen operates as a permeable membrane that allows binary oppositions (presence/absence, past/future, etc.) to conjoin and stabilise in a coherent experience that, above all, we must believe.

Was cinema the most successful mind-altering technology of the 20th Century?

Beyond the script; machine perception

How else could we resolve the simultaneity of fragmentation and unification, or of imagination and perception, except through the impulses of desire, and the things it propels us to do? Perhaps we are straying beyond the bounds of what it possible to describe with scientific maps, but given that paradoxes are by their nature irresolvable, cognitive science needs to furnish some new explanation of its object if it is not to evaporate, as some suggest it might. The new models of the mind that have recently been proposed cannot be regarded as entirely the consequence of silicon-chip technology. The insights into language and myth, alluded to above, have been formative in determining the way we think about ourselves as both ubiquitous and amputated, dispersed and. localised. These conceptions, no less radical than the *Surrealist Manifesto* or Marinetti's exhortation to "burn the museums", has opened a window for expanded thought which embraces contradiction as a means to unification and recognises the essential continuity of all reality (see note 4).

The inverse hollow

We have suggested that human desire may be a consequence of the deep dissatisfaction we have with the fragmentation and contradictions of consciousness that is implied, at least partially, in the processes of language (which itself must now be seen as a technology). Given this, the attraction of objects and ideas that minimise the fragmentation in our perception of the world seem obvious. Many apparatuses and technologies already exist as desiring machines, the cinema prominent amongst them, which function to induce a sense of coherence and stability in our mind through various narrative and compositional strategies. In the postdigital continuity between language and technology, our desires are both satisfied and generated as we reconcile the fracturing that appears to be a pre-requisite of apprehending meaning.

The attraction of objects that minimise fragmentation seem obvious

The attraction of objects that minimise fragmentation seem obvious

Section Five

PROPOSITION:
The distinctions between the words 'human' and 'machine', and between humans and machines, lose validity as meanings mutate in response to imagination and desire.

Could machines know?

Could a machine think?...Could it be in pain? - Well, is the human body to be called such a machine? It surely comes as close as possible to being such a machine.

Ludwig Wittgenstein, *Philosophical Investigations*

Intelligent machines in historical and academic context

The desire to invest matter with vitality is evident in ancient Greek legends such as *Pygmalion* in which clay figures are animated. The mechanical automata of later centuries simulated the functions of humans and animals so well that Descartes was inclined to regard animals, with the exception of humans, as machines. Philosophical speculations about the mind from the mid-seventeenth century onwards, notably in Descartes and Leibniz, suggested to Hobbes that human reason was itself computable. More recently it has been thought that computing machines have presented a serious possibility of replicating human thought. In the 1960s electronic computers capable of rapid symbol manipulation fuelled the hope of defining a formal reasoning system which could be subject to machine processing. The very idea of an intelligent machine, however, has met resistance from philosophers who fundamentally disagree with the concept of the human mind as a symbol processor. Other objections regard organic matter as having certain properties that are specific to the organism and not reproducible in the hardware of a computer. Despite such objections, recent decades have seen a huge investment in the possibility of Artificial Intelligence.

The rise of the AI paradigm

As with many technological initiatives, the Second World War provided a significant stimulus to intensive research into intelligent machines. The demand for cipher and decoding machines, 'intelligent' weapons and remote surveillance was unparalleled and Alan Turing's electromagnetic decoders formed part of the tapestry of devices from which the von Neumann computer subsequently emerged. The assumption that human intelligence might be replicated in mechanical devices drew strength from advances made in calculating the complex mathematical equations used in artillery. In 1946, when the ENIAC computer was completed and publicly displayed, the trajectory of a shell was calculated in less time than the missile took to reach the target. This demonstration suggested that the calculating machine might be used to extend human 'brain-power' beyond the constraints imposed by nature. We had entered the virtual age and with it came a link between machines and a disembodied intelligence. This launched research programmes based upon the assumption that artificial intelligence was an achievable goal. Henceforth, advocates of strong AI proposed that intelligence was a formal system of symbol manipulation that was independent of the physical body and reproducible in a machine.

"More human than human"

If anyone had any doubts about the extent of the claims made by advocates of strong AI then this passage from Daniel Dennett should dispel them:

> If the self is "just" the Center of Narrative Gravity, and if all the phenomena of human consciousness are explicable as "just" the activities of a virtual machine realised in the astronomically adjustable connections of a human brain, then, in principle, a suitably "programmed" robot, with a silicon-based computer brain, would be conscious, would have a self. More aptly, there would be a conscious self whose body was the robot and whose brain was the computer. The implication of my theory strikes some people as obvious and unobjectionable. "Of course we're machines! We're just very, very complicated, evolved machines made of organic molecules instead of metal and silicon, and we are conscious so there can be conscious machines — us." For these readers, this implication was a foregone conclusion.

Daniel Dennett, *Consciousness Explained*

It is not that we are picking on Dennett — he is merely typical of a number of writers on the topic. This passage is included, however, because it symptomatically collapses a theory into a wish. Whilst the weaknesses of the strong AI case are simple and obvious, the overwhelming desire for synthetic consciousness obscures them. Ironically, those who desire robotic intelligence will probably get their satisfaction (the energy of human desire laughs at logical argument) by forcing the mechanical paradigm to its limits.

> There is no other way of conceiving the indestructibility of unconscious desire — in the absence of a need which, when forbidden satisfaction, does not sicken and die, even if it means destruction of the organism itself. It is in a memory, comparable to what is called by that name in our modern thinking-machines (which are in turn based on an electronic realisation of the composition of signification), it is in this sort of memory that is found the chain that insists on reproducing itself in the transference, and which is the chain of dead desire.

Jacques Lacan, *Écrits*

Intelligent machines defined

Interest in computing machines, with their potential for simulating (even amplifying) the kinds of cognitive activity formerly thought to be the sole prerogative of humans, was not confined to computer scientists. The philosophical debates surrounding AI shortly after the war threatened to stifle investigations that were, at that time, clearly poised to take advantage of massive military research budgets. To overcome this barrier to further practical investigation the mathematician Alan Turing proposed a test for verifying the intelligence of a machine that is still considered valuable. The Turing test, significantly first published *Mind* (the journal of psychology and philosophy) in 1950 relies upon the human (tester's) perception of a machine's response to unrehearsed questions. However, the primary purpose of the test was to provide a benchmark that cleared the ground for expensive speculation in the computational sciences. Subsequently, the more limited goal of imitation replaced that of replication and human intelligence was confused with conversational ability. In making his prediction for the future of intelligent machines, Turing re-affirmed a belief in the indeterminant nature of the inquiry into machine intelligence:

> Nevertheless I believe that at the end of the century the use of words and general educated opinion will have altered so much that that one will be able to speak of machines thinking without expecting to be contradicted. I believe further that no useful purpose is served by concealing these beliefs. The popular view that scientists proceed from well-established fact to well-established fact, never being influenced by unproved conjecture, is quite mistaken. Provided it is made clear which are proved facts and which are conjectures, no harm can result. Conjecture is of great importance since they suggest useful lines of research.

> Alan Turing. *Computing Machinery and Intelligence.*

By addressing the readers of *Mind*, Turing signalled a specific area of research — the modelling of the mind in relation to a technological environment. He also suggests a methodology that reflected his understanding of the practice of scientific research as a combination of imagination and verification. An agreed objective test of an intelligent machine, he believed, freed research from a logically coherent pathway to allow the interventions of imaginative thought. In this move Turing laid the foundations for an interdisciplinary study which subsequently formed around the topic of Artificial Intelligence research.

Philosophical challenges set aside

Prior to Turing's rather limited definition there existed no stable, objective criteria against which progress in machine intelligence could be judged since prevailing models of human thought (from science and philosophy) were in fierce competition rather than gentlemanly agreement. However, as a consequence of Turing's intervention, early experimental work into artificial intelligence was, to a large extent, freed from inhibitive philosophical scrutiny. Practical projects could proceed to try and show that mechanical models using recurring numerical series were able to manifest intelligence. But the motives for agreeing on one model over another may have had less to do with epistemological integrity than with funding criteria and the necessity for consensus.

Practical AI and the rhetoric of funding

Artificial Intelligence was a subject that thrived at the interface of philosophy and mathematics. But such research is not always supported entirely for purely intellectual reasons. US military interest in various projects during the early 1960s and the 1970s suddenly made available huge research budgets administered by the Defence Department of Advanced Research Projects Agency. Amongst competing models, a computational approach to AI based on the idea that thinking could be rendered in algorithms predominated at the expense of the organically based connectionist models epitomised by Frank Rosenblatt's invention of the 'Perceptron'. The appeal of the algorithmic approach to military minds was irresistible, no doubt partly influenced by the wartime success of Norbert Weiner's air-artillery computations. Where the connectionists spoke of concepts and biological soup, computationalists spoke of routines, heuristic algorithms and outcomes. It was not until the mid-1980s that the conceptual and practical limitations of the computational approach were widely recognised. Then there was

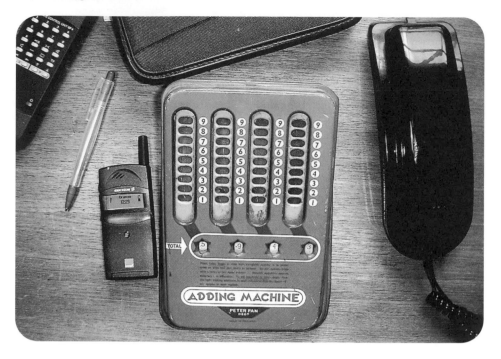

Philosophical challenges set aside

a re-flowering of the Perceptron in a potent cocktail of behaviourism and biology that enabled a fragmented model of the mind to be reconciled with Holism. The resurrection of this approach in the 1980s was under the rubric of Parallel Distributed Processing (PDP). Two considerations argued strongly in this direction. In the first place, serial-processing scales badly since it requires bigger and bigger machines with faster and faster processors to cope with more sophisticated cognitive processes. This might be called a technological limitation. Secondly, what constitutes intelligence is not static — in cinematic terms Cecil Hepworth's clever dog 'Rover', who found the baby in the attic, had to give way to 'Rin Tin Tin' (a much smarter dog altogether) and then to 'Lassie' and to 'Beethoven', and so on. Each successive animal requires an upgrade as explanation quickly catches up with mystification. AI had to learn the lesson of all scientific research — everything we think we know is probably wrong — and adapt quickly to changed circumstances in order to keep funding interest.

Humans will go to extraordinary lengths to distribute their intelligence in other matter

The model becomes the object it models

The methodology of reducing complex systems in order to understand their workings is common to scientific and mathematical procedure. In particular, the reduction of a problem to a sequential progression of binary absolutes (on/off) is the essential task of the computer programmer. Yet, as with art, the skill of the practitioner may mask the inadequacies of the medium. Conceiving processes in binary terms is necessarily planar and sequential, despite the fact that the mapping of alternative routes may be more conveniently represented in a three-dimensional schema, or with simultaneity. Despite brilliant simulations, what is lost when the mind is represented as a binary operation is the deep associative structure of emotion and feeling which is replaced by a linear, logical network. The understandable tendency, therefore, is to disregard or underplay those aspects of mental behaviour that defy logical organisation, such as irrationality or spontaneity. As a consequence it seems that computers, however advanced, are rarely characterised as anything other than consistent and impartial (even if this means they act stupid).

77

Models are limited

Our habitual characterisations of computers have their roots deep in the optimism of enlightenment, from Descartes to Lévi-Strauss. The Structuralist impetus to AI research proposes a powerful psychology in which the resistant depths of the Freudian mind are replaced by a complex (and perhaps equally unknowable) surface web that logically processes information to produce a response. We necessarily adopt a limited model in order to make progress. But in the absence of anything better the limited model inadvertently comes to stand in for, or take the place of, the object it means to represent.

Tensions in the symbolic paradigm

The mind model that emerges from the same camp as structural linguistics and anthropology, behavioural psychology and symbolic logic has been exposed as deeply flawed due, not least, to a number of conflicting imperatives that created unsustainable tensions amongst researchers and apologists. These tensions arose mainly as a consequence of trying to map one object (the mind) onto another incompatible object (the computer programme) in the hope of forcing continuity between them. In doing so researchers failed to see the continuity that already existed between mind and machine, a continuity that only becomes apparent when the mind is freed from the brain. Ultimately, their research could not be reconciled with perceived reality. It undermined its own foundation in that it was trying to model a model (using a flawed model in each case) — a lesson which has particular bearing on contemporary theory, especially in the humanities.

The importance of metaphor for understanding

As we now know, little practical progress was made in replicating human intelligence, but if nothing else, AI research has brought to the fore a number of fresh insights into crucial philosophical problems. Recent work in the history of science has shown the extent to which evidence achieves professional currency through the acceptance of agreed metaphors. The paradigmatic example often cited is that of Friedrich von Kekulé who claimed to have imagined the structure of

Benzene. He recounts how in the 1860s he dreamt of a radical molecular structure for carbon that subsequently laid the theoretical foundations for carbon chemistry. Kekulé's model prevailed for more than a century before it was exposed as flawed. However, since it allowed agreement on some puzzling aspects of organic chemistry it was accepted and prepared the way for the double-helix metaphor that is currently used to explain the structure of DNA. This metaphor was itself a product of the rhetoric of funding as Watson and Crick developed the double helix as a compromise between a number of other prevailing representations, some of which had equal scientific credentials. What these developments demonstrate is the powerful role of imagination in shaping our conception of the objective material universe and the contingency of models that are based on our ability to recognise what we think we've discovered.

Metaphors change with the dominant technology

Metaphors change with the dominant technology

The recognition that models of reality might be subject to shifts in intellectual vogue, or contingencies of funding, casts doubt on any claims they may make for universal applicability. To explain the boundless complexity of the universe we are simply offered the most complex mechanisms known at the time. What model of human intelligence, for example, is implied if we find repugnant the idea that a car might recognise its own reflection? Since we do not have the scope to express what cannot be imagined we are restricted to that which is expressible by association (modelling). In refuting the computer metaphor of the mind John Searle observes:

> ... in my childhood, we were always told that the brain functioned like a telephone switchboard. The brain, in short, was a cross-bar system. (What else could it be?) I was fascinated to find that Sherrington compared the brain to a telegraph system. Freud frequently uses electromagnetic and hydraulic analogies in discussing the brain. Leibniz compared the function of the brain to the functioning of a mill; I am even told that certain Greek thinkers thought that the brain functioned like a catapult.

These technological representations of the brain are for Searle an unconscious masking of the lack of a neurobiological explanation. However a neurobiological model would, for some people, be regarded as yet another metaphor by which to order the available facts. What the history of AI, from Turing to Searle, makes apparent is that new machines frequently invite different, sometimes unprecedented, conceptions of what distinguishes a human. Yet each new conception betrays the same desire to integrate the human and the machine, or at least provide a mechanical description of human faculties. Though machines are currently not considered capable of simulating human intelligence the possibility of extra-organic thought (thought outside the brain) has forced us to rethink some deeply ingrained assumptions about the delimitations of the material world

Transcending machines

What is at stake in the question of machine intelligence is not so much what machines will or will not do in the future, but the very definition of humans as not being machines. We have always been able to salvage 'humanity' from the erosive threat of technology by recourse to an irreducible spirit — appealing to a perceived complexity in our nature which lies beyond comprehension. Yet spirit is also a quality we easily ascribe to things apparently far less complex than ourselves. We can readily sense an intangible geist in animals, carvings, paintings, dwellings and all manners of inanimate objects. We know there is a institutional reluctance to ascribe spirit to machines when at a personal level it is commonplace, for example the cars we made out in, and the planes flown by pioneers; in short, machines with more than technical meaning. It seems inevitable that eventually we will invest spirit in the very systems we design to think for us, to be in the world for us. Such machines (mechanical or biological) will appear to have transcended the demarcation that, until now, has distinguished us from them.

What happens to a thing if it no longer conforms to its own definition?

Let us say the definition of 'machine' is an arrangement of components that automatically performs prescribed functions. Even the most sophisticated automaton would normally be defined by having certain qualities such as servility, a lack of self-awareness and the status of controlled subject. What happens to this definition when a machine no longer displays all or some of those qualities? What if machines developed to an extent where they were no longer considered lacking in intelligence, or no longer subject to our direct control (like some PC operating ssytems)? At this point there are two things that might result — either the machine becomes something other than a machine, or the definition 'machine' is altered to accommodate the new circumstances. In either case the machine as an object will have transcended its prior status and become something more than the sum of its components.

Can machines transcend the mechanical?

The machine that displays characteristics of human consciousness is neither a machine nor a human. It has transcended either classification and becomes a transcendent machine.

Transcendent machines

At the beginning of the twenty first century we seem to be faced with some conflicting trends in the development of ideas. On the one hand, there is growing sympathy for those philosophies that reject mechanistic and materialistic views of the Universe. Among others, the surging interest in Shamanism, Witchcraft, Paganism, Buddhism and Spiritualism testifies to this. The tendency of these modes of understanding is to (re)embrace notions of emergence, holism and cosmic union in contrast to the divisive, categorical and empirical modes that have dominated Western thought for over two thousand years. Yet, on the other hand, the anticipated consequence of research in fields such as strong AI, Neural Networking, Artificial Life, etc. is that we will model — even replicate — indefinable, 'transcendent' states of human existence such as consciousness and Being, using machines based on fixed computational logic and binary architecture. Those who see this as a conflict between Mysticism and Science, or as paradoxical, forget that we are all looking at the same world and such oppositions are now unhelpful and are inevitably being transcended.

Are we ready for what we want?

In the *Post-Human Condition* it was argued that a convincing artificial intelligence cannot be 'programmed in' but will almost certainly 'evolve' or 'grow' given the right conditions. However, a mindset that disavows the continuity between science and spirit, mind and machine, technological and human, will be completely unprepared for the consequences of such a development. Contemporary legal, political and moral systems are still rooted firmly in the nineteenth century (it is only recently that the British Parliament has considered archiving its work electronically rather than on vellum) and there is evidence of a struggle to accommodate contemporary technology into the social order. For example, recent controversies have led to anything from amusement to moral panic since they insist on new levels of intimacy between disassociated spheres of human existence. We need only think of a few contemporary examples:

• Dolly the 'cloned' sheep (uniqueness/replication)
• Foetal gender selection (randomness/predictability)
• Genetically modified foods (natural/synthetic)
• Patented human DNA sequences (private/public)
• Sperm freezing and post-mortem fertilisation (life/death)
• Commercial or governmental use of data (freedom/control)
• Pornography on the internet (desire/taboo)
• Freedom of speech online (anarchy/democracy)
• Transgenic organ transplant (animal/human)

This list is evidence of the widening gap between the dynamic growth in technical applications and the increasing inadequacy of institutional legitimacy. Such moral and ethical turbulence, however, will seem minor in comparison to what might be in store for us if some researchers are to be believed. The emerging prospect of 'machines with souls' or machines that can talk to us, seduce us or deceive us, poses an altogether new layer of moral conundrums that have only really been explored in science-fiction literature. Given the huge amount of resources invested world-wide in medical, scientific, military and bio-chemical research, more technologies will certainly emerge to cause further concern as the apparatus for making sense of the world is overtaken by the machines we invent to satisfy desire.

Science generates anxiety as it resolves desire

Ethical anxieties appear to be an inevitable outcome of modern science and technology. In 1818 Mary Shelley posed the dilemma about the status of beings produced by human scientific ingenuity rather than 'natural' processes in the novel *Frankenstein*. It is no coincidence that this book is still invoked with reference to the perceived horrors of cloning and genetic modification. In bleaker moments it must seem that the abrogation of our own unique being (which, granted, is also nothing new) has been the avowed aim of institutionally sanctioned (and publicly funded) research for two centuries or more. Given the hope invested in such projects, it is unfortunate that science is only able to resolve anxieties at about the same rate that it generates them. Hence, we are caught on a hook, swinging between the perpetual antagonists of desire and satisfaction.

The redundant genius

The desire to produce a machine more ingenious than ourselves both defines the human as supreme and subjects it to the terror of redundancy. This self-flattery demands nothing less than the extension of our own intelligence beyond our comprehension. Proceeding on the basis that only problems for which there is demonstrable solution will attract funding, we torment ourselves with schemes like artificial intelligence. AI is essentially a problem arising from a pre-defined solution, the answer being already concealed in the expression of the conundrum like some insane crossword puzzle where the clues have to be derived from the answers. However this is not to completely dismiss AI research. It has at least defocused the sharp boundaries that used to confine intelligence to the brain. Implicit in the solution to its own problem is the assumption that thought can exist outside of the mind. Artificial intelligence, therefore, is already 'real' intelligence inasmuch as it embodies our intelligence in a coded form. To insist on it as a mode of intelligence separate from, or outside us, is simply to reassert the redundant divide between human and machine, mind and body. Only by delimiting intelligence can it putatively exist outside the body as a technology. The consequent moral panic is as inevitable as scientific change. Although terms such as 'real' and 'artificial' have linguistic significance, this should not mask the seamless continuity between these two domains, both of which are consequences of human imagination.

Humachines

Humachines

This section has outlined the arguments against the reductive description of 'mind as machine' whilst situating it historically in the accretion of scientific metaphors about the operation of our consciousness. In this respect, the binary paradigm is but the latest in a series of analogies through which humans have tried to neuter the mystery of sentience. The postdigital membrane reiterates the idea of a symbiosis between desiring humans and machines — a thick theory in which technology and ideas are led by each other in a sort of snake dance where each claims the high ground. In fact, as has been shown, theory and hardware have always been intimately entwined in the shifting epistemic fashions and scientific assertions that guide us. To claim priority for any one over another is to miss the point as recklessly as James Cole's time travelling does in *Twelve Monkeys*. We will end up in the trenches of the First World War when we wished we were in Philadelphia.

Intelligence and energy

Intelligence and energy

Section Six

PROPOSITION:
The practice of art consists in arranging matter so as to add (extra) significance to it.

A plate of food scraps

If I distrust my memory...I am able to supplement and guarantee its working by making a note in writing. In that case the surface upon which the note is preserved, the pocket book or sheet of paper, is as it were a materialized portion of my mnemic [sic] apparatus, which I otherwise carry about with me invisible.

Sigmund Freud, *A Note Upon the 'Mystic Writing-Pad'*

Rock art to Web art: Projecting the imagination

We are free to speculate about the function of images applied to cave walls and rock surfaces by distant cultures. Inevitably, we will be inclined to interpret them from the intellectual convenience of our present historical perspective, casting the light and shadows from our own candles and reflecting our current sensibilities. Whatever speculative interpretations we now place on these images there are a number of reasonably secure claims that we can make. First, these works are the product of human activity. Secondly, the beings that made them enjoyed a level of intelligence that we would recognise in ourselves (perhaps greater intelligence than we had previously imagined if recent theories about astronomical maps in the caves of Lascaux are to be credited). Thirdly, the paintings and carvings were not random deeds of mutilation or defacement but intentional creative acts arising from motives peculiar to the circumstances in which they were conceived. Whatever moved those ancient 'artists' is still the subject of academic research and open to debate (even if we do eventually make a proper time machine for James Cole there is no guarantee he will understand pre-historic motives). What these images do have in common with contemporary experience is that they demonstrate the impulse to modify what is around us through a combination of thought and action, collapsed into a single coherent process. It seems we have rarely been content to leave the world unchanged, perhaps believing that our survival depended on changing it. More than any other species we change our surroundings to suit our needs and we can look to any urban space to see there is very little that has not been altered by human intervention. Whilst this is not a revelation, it is a fact of foundational importance that can go unnoticed. Looking into an ancient cave reminds us of our persistent desire to modify the environment according to our imagination.

Palaeolithic Rock Painting

What we recognise in those rock artists of the Palaeolithic era is the impulse to integrate our mental and physical space, crossing that perceived divide through the agency of imagination. It may be that a sophisticated language faculty is the defining characteristic of humanity, but this is only one example of our desire to draw the world into ourselves and ourselves onto the world. In this sense there is a direct genealogical trail stretching across human history that is expressed today when people upload pictures onto their homepages. Such acts are examples of the attempt to imprint our personality (the sum of mental and physical activity) in an enduring medium that becomes subsequently accessible. In the process there is relay of imaginative impulses that pass from one side the human membrane to the other, reminding us of the indivisibility of the two spheres of human existence (internal and external).

Significant matter

It is often the case that art, regarded as imaginative entelechy, transcends the internal and external spheres as it absorbs the quotidian, the concrete and the transcendental. An example exists in the *Annunciation with St. Emidus* by Carlo Crivelli hanging in the National Gallery in London. This painting has been there since 1864 when it was presented by Lord Taunton in a philanthropic act which linked charity with the material value of art. It was an apt choice since the painting itself mixes the earthly with the divine. Today it has a different purpose, surrounded as it is by art in the National Gallery of a European capital. But during 1486 it was intended for the church of Saintissima Annunsiata in the town of Ascoli Piceno and since then its meaning has been inflected more by other paintings and political and economic events than the whispered confessions of the faithful in a provincial church. It is essentially a highly accomplished, carefully constructed statement, designed to be read and understood in conjunction with the New Testament, candles and chants.

Art as a connection between discontinuities

In a *mise-en-scène* more befitting James Cole's adventures, the carpet on the balcony in the painting connects the past and the present by depicting the contemporary practice of hanging carpets on Good Friday. Inside the house the symbols of a good woman abound, her house is tidy, the plates stacked, bed made, cushions ordered and so on. In all respects it is a purposeful composition which celebrates contemporary technique by including two trompe l'oeil fruits attached to the surface — an apple to signify the fall and a gourd to symbolise the husk of the human body. It is because of the evidence of this careful design that we are surprised by the Angel's partial interest in the Virgin and the utter indifference shown to her by the patron Saint of the town, St. Emidus (perhaps it is because she has yet to be famous). This sweet-faced man is holding a model of the town of Ascoli Piceno, the very town that provides the setting for the annunciation. We know that on March 25th 1482 news reached the town that the Pope had granted a certain measure of self government to Ascoli Piceno and that the painting was commissioned to celebrate the two annunciations, that of the angel and that of the Pope. Here past and present, material and ethereal, private and public, local and universal are fused in a masterful exposition of contemporary art practice. Oil, pigment and cloth are skilfully arranged so as to semi-permanently encode the imaginative excitement of these events. Stylistically so distant from the cave paintings of Lascaux, this image merely retreads the desire inscribed on the walls in previous ages and in many distant places.

The functions of artistic expression

In trying to account for the function of artistic expression as a near-universal phenomenon we can cite a wide range of possible motives: serving as an indicator of status or class, focusing religious sentiment, tribal identification, supplying a commercial market, decorative adornment, recording information, passing on knowledge, and so on. Whilst it would be simplistic to attribute something as complex as artistic production to a single determinant it may be that the various functions cited have something in common, or at least an inter-related significance; that is, the compulsion to manipulate and organise matter such that it evokes something other than it is. This compulsion can puzzle even accomplished artists:

> It seems strange to me that someone thought of making marble statues. I understand how you could see something in the root of a tree, a crack in the wall, in an eroded stone

or pebble. But marble? It comes off in blocks and doesn't evoke any image. It does not inspire. How could Michelangelo have seen his David in a block of marble? Man began to make images only because he discovered them nearly formed around him, already within reach. He saw them in a bone, in the bumps of a cave, in a piece of wood. One form suggested a woman to him, another a buffalo, still another the head of a monster. (Pablo Picasso)

Brassaï, *Conversations with Picasso*

The factors of recognition and resemblance

The observation of small babies seems to confirm the immediacy with which we are able to re-cognise the content of a representation (a toy bear) and suggests such recognition is a primary cognitive faculty. Indeed, it may be so primary that it is prior to the faculties of comprehension or analysis, and therefore impervious to scrutiny. In which case, the search for its origins could only be a matter for informed speculation and theoretical debate. However, if we have any intellectual investment in the idea of evolution then we would be inclined to say organisms use their sensory apparatus to survive and gain reproductive advantage. For reasons that do not require elaboration here, the faculty of recognition (Latin: to know again) would have been of significant benefit to the development of our species. It has become clear through a number of psychological and physiological experiments that recognition is a significant part of sensory processing, that is, the correlation of sensations with learned or innate perceptions. It would seem sensible that such a complex would respond in a similar way to a stimulus correlating to a given object which was not actually the object perceived. For example, we may see a black bag on a chair and think it is a cat — a case of mis-recognition. Nonetheless, for the time that we perceive it as a cat it is a cat, as far as we are concerned, and if we are frightened by cats we will be genuinely afraid of the bag until we see it for what it actually is. Likewise, through similar action of stimuli and perception, we can wilfully perceive equivalence in objects that share perceptual qualities — a mountain range might suggest a face in profile or a tree stump may look like an old man — a case of resemblance. The critical difference between a mis-recognition and a resemblance is the mis-recognised object starts by apparently being one thing

91

and then becomes (almost irredeemably) another — the two perceptions are mutually exclusive and consecutive. Meanwhile, the resembling object can both be itself and be like the thing it resembles — the two perceptions are mutually exclusive but oscillatory.

The factor of representation

The processes of mis-recognition and resemblance are necessary components of representation — the phenomenon whereby matter or energy is intentionally arranged so that we perceive something that is absent. Representation demands that we (almost permanently) mis-recognise the material from which the representation is made because the material resembles so strongly something other than itself. With visual representation, as for example in a statue, an object resembles more closely something it is not (a figure or animal) than something it is (stone or wood). In this instance the materiality of the sculptural medium is subordinate (by repression and perceptual delay) to the representational significance of the figure — the appearance of what is absent over-rides the actuality of what is present. In the drawing below we recognise a human head as a consequence of an arrangement of ink marks that are all but invisible at the moment of perception.

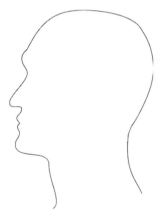

In this drawing the factors of mis-recognition and resemblance collectively inspire the representation of an 'absent' head (which is also a mark on the page)

The operation of the mark

Whilst not all representations are made with marks (other technologies such as electron streams are available) mark-making, along with carving, is undoubtedly one of the primordial ways of arranging matter so as to endow it with extra significance. In being so arranged, the representational mark renders the conscious impulse that created it, an impulse that can then be relayed back to a viewer (perhaps intact) when perceived and recognised. Hereupon the mark itself virtually disappears, or at least becomes transparent, inasmuch as we apparently see it for what it purports to be rather than what it actually is (as in the picture above). We are culturally bound to mis-recognise the mark for what it resembles, to imagine an absent object or idea in its place. Yet, as with the operation of the Mystic Writing-Pad, a significant trace of the mark itself (or the object being represented) is never entirely absent from the picture.

The envaluation of the mark

The imaginative operation of mis-recognition and resemblance, as they conspire to produce representations, normally requires that the technology of the mark remain largely transparent. If we read the drawing above as merely a squiggly line of ink then we would be seeing only the mark itself, presented through the technology of print. Yet also apparent within the technology of marks are assumptions about the cultural value of the imaginative impulses they relay: there are varying qualities of print. Dry mark-making invariably involves a soft smooth material — lead, graphite or wax — being drawn over a relatively hard, textured material such as paper or canvas. This kind of mark is the consequence of some of the soft material being abraded, and the detritus from the process becoming absorbed in the irregularities of the surface of the support. Erasure is restricted by the absorbency of the support, which normally means subsequent clean mark-making is rendered impossible. The blackboard works in a somewhat different way. A hard smooth surface supports the grains of a compressed dust that are deposited when the pressure on the chalk, as it is dragged across the surface, exceeds the pressure exerted by the binder that holds the grains together. The adhesion of these grains to the surface of the board is a consequence of attraction at the molecular level and residual moisture in the chalk, which draws water from the atmosphere. Since no deep absorption occurs on the surface the dust particles can be relatively easily removed. By sacrificing a degree of permanence then, erasure and correction are, to all intents and purposes, infinite. Such fluidity encourages a type of imaginative improvisation that is able to erase its immediate past without regard its own position in the future. The radical feature of the blackboard was that its use acknowledged the impermanence of some marks and what they represented, implying therefore, the possibility of imaginative flux, since the marks that embodied imagination were, themselves, reconfigurable. The device has frequently been used as a visual metaphor for abstract thought in motion — Albert Einstein was famously pictured with a blackboard full of scribbled equations. For most people, of course, his mathematical marks remain almost entirely opaque, which only serves to reinforce their fleeting purpose. The mark that is absorbed into the support, by contrast, has the virtue of permanence which, thereafter, permits it, along with the imaginative activity it represents, to enter the temporal order of matter and gain an historical (and often legal) validity. As such it resists becoming ephemeral, to the satisfaction of those who deny the present. This materiality of the mark reflects the general trend that privileges rigidity over fluidity, inscription over improvisation. This is further demonstrated by our consignment of the blackboard to the heritage-park and our huge investment in permanent archives and galleries.

Light technology

Up until the mid-nineteenth century, viewing representations, such as paintings, by daylight would have been the exception rather than the rule. From cave-painting to church icons to courtly galleries, flames were used in one form or another as artificial light to illuminate the marks that rendered the representations. Yet, paradoxically, the least visible constituent of the imaginative process of visual representation is light itself. In the case of artificial light, the radicalism of the candle is easily overlooked. Prior to the candle, light was generated by burning sticks. The best were those impregnated with resins and these burned to ash as they gave illumination. In a startling inversion of the dynamic, the oil lamp, burning vegetable oils or tallow (animal fat) used a crude wick which was not consumed in the production of light, the fuel was exploited whilst the conduit was everlasting.

Both means were somewhat unpredictable, inefficient, and generated fumes which limited their applications. The candle differs from the tallow lamp in an important and radical way. Although both draw inflammable oils into a flame by capillary action, the candle uses the heat of the flame to liquefy the oil and this renders it safe and portable, with the added advantage that latent light could be stored for long periods in a compact and relatively stable form. In 1783 a new kind of wick was developed which fed oxygen into the centre of the flame and enabled all the carbons to be consumed in the combustion processes giving increased luminance and less fumes. An ingenious refinement was introduced by the addition of a filament that contracted on burning causing the wick to curve and its tip to touch the hottest part of the flame and burn away rather like the resinous stick did. But for this arrangement the flame would become smoky as the wick became longer. Most of all, however, the candle was not just a 'super stick' in which the fuel to waste (noise to message) ratio was spectacularly reversed, it was an efficient source of light that could be mass produced and distributed within the circuitry of an emerging industrialised society. Candles quickly became essential paraphernalia to the activity of both making and seeing art. Yet, in the same way that the mark becomes transparent at the moment it refers to something other than itself, candles remained largely invisible as attention focused on the objects they illuminated rather than the imagination embodied in the remarkable technology of light.

Artificial day and extended humans

On a wider secular scale, industrialised light can be seen to have had enormous social effects. Leisure, such as it was, could be extended, there were greater possibilities for travel, and education could take place in the hours when work ceased. There was even more time to devise new forms of pleasure including making and consuming representations (such as art). The perceptual limits of the human eye most immediately effected the length of the working day and the economic survival of the individual worker, and thereby the success of the community. But it also effected memory. In an essentially non-literary society the storage and retrieval of data relied upon sophisticated mnemonic devices including folk tales and poetry. Cheap (as candles eventually were) and reliable forms of illumination extended literacy and calculation beyond the leisured and the institutional whilst simultaneously widening the franchise of these groups. Writing becomes immanent and less regulated, and the candle, in this respect becomes part of the dispositif of an information storage and retrieval system. The cheap candle was not only part of a new industrial and intellectual age but also a contributor to a new self consciousness in respect of history and prospects. Which reminds us that the writing pad, the painting, the mark-making devices, the candle and subsequent technologies of representation affect the extent and location of the human and our imagination. The technologies of the representational image, combined with artificial light, opened up new spaces that extended the possibilities of human being, and stimulating new desires (see note 5).

Art as technology

The technologies of representation, at their most basic, consist in arranging matter or energy so as to invoke something that is absent. The 'art' or skill of this practice consists in the way in which the artist is able to manipulate the material to invoke what is absent, i.e. it can be done either well, poorly, brilliantly, superficially, etc. This is the traditional and most widespread sense in which the term 'art' is applied, which would link it directly to the Greek root of 'technology' — *techne* — meaning art or skill; this is how art has been understood through most of classical art history. Art, therefore, is a technology we have developed that allows us to invoke the presence of things in their absence by skilfully adding extra significance to matter or energy.

> Art always makes use of a material, technical, tool-like device, of an appliance, a 'machine', and does so so openly that this indirectness and materialism of the means of expression can even be described as one of its most essential characteristics. Art is perhaps altogether the most sensual, the most sensuous 'expression' of the human spirit, and already bound as such to something concrete outside itself, to a technique, to an instrument, no matter whether this instrument is a weaver's loom or a weaving machine, a paint brush or camera, a violin or — to mention something really frightful — a cinema organ.

> Arnold Hauser, *The Social History of Art*

New technology of art

The description of artistic practice as the skilful addition of significance to matter may seem valid for the largely depictive art works produced up until the beginning of the twentieth century, i.e. those works associated with the great museums of the industrialised world. But we also know that, in addition to the manipulation of paint, stone, metal and graphite, etc., there are a number of apparently new dimensions to the practice of art as we understand it today. The application of the word has become much more diffuse in recent history and the varieties of expression more diverse, for example, activities now considered as artistic include self-mutilation, bioengineering and telecommunications. Although few people today can agree on the meaning or function of art, it does not follow that it never had either. Given that art is no longer regarded as having a specific object or mode of practice and that works of art are less frequently rendered in paint or brass, it might be thought that the theory of art described here would be deficient in relation to contemporary practice (with its apparent rejection of many traditional art making mediums in favour of 'new media'). On the contrary, whilst the materials employed have changed radically, the essential operation of representation remains the same. Instead of arranging rocks, dye on stone walls, burnt earth bound in oil, or ground up gemstones in egg yolk, we are able to manipulate silver crystals and dots of ink, electrons and diodes to invoke things other than themselves. From computer screens to IMAX cinema, whether still, moving or 3-dimensional, whether printed or projected, artists (along with others using the media) who make things today use the same operative principle to conjure up absent things or ideas as the Palaeolithic artist used.

Abstraction and non-representation

One could object that this description of art might apply to 'representative' or mimetic works (having a recognisable subject) but not to those that apparently speak only of themselves. Paintings of High Abstraction such as those by Jackson Pollock and Barnett Newman are often cited as examples of an art that makes no reference other than to its own surface. Modernist abstract art was supposed to be no more than itself. It was not meant to 're-present' things by suggesting the presence of something that is absent. Putting aside whether or not the total evacuation of external reference is ever attainable, abstract art remains consistent with our description inasmuch as these works also require the skilled arrangement of material to manifest their shapes and forms and add extra significance to matter. The paint from the tin has to be organised as squares, lines, squiggles, etc. on a canvas in order that the desired significance is produced. The subsequent meaning attached to the shapes and colours then becomes the subject of interpretation. Despite their apparent intention, abstract works never escape representation but, at best, sublimate it in ideas (or modalities as Clement Greenberg had it) rather than objects — ideas like Modernism, spiritual enlightenment or dramatic personal catharsis. In other words, abstract images are also made from material that has had extra significance added. Something of Pollock's emotional condition is present in his works, even if he is absent.

Matter gains extra significance when it is arranged

Art without skill

A major exception to our argument about the skilled arrangement of matter would seem to be the practice that designates pre-existing found objects as works of art 'in themselves'. Early exponents of this approach in the Western Art tradition such as Marcel Duchamp (following Pablo Picasso and Georges Braque) and the Dadaists, have been followed to this day by countless others. While 'found art' pieces require some manipulation and arrangement of the material (after all, the objects have to be selected and mounted) it is not generally intended that the chosen material stands for anything other than itself. Duchamp's *Fountain*, therefore, was simply a urinal chosen, so he claimed, for its "aesthetic neutrality". Whilst this was a significant advance in terms of the ideology of art, such simplicity of interpretation was short-lived (in fact, it was never so simple). Almost immediately after its submission to the 1917 *Society of Independent Artists Show* in New York, the object became encrusted with additional connotations — such is the power of art to bestow significance on any matter anointed by its name. It was even claimed that its shape evoked a seated Buddha and it duly acquired the nick-name 'Buddha of the Bathroom'. As a consequence, the matter of the urinal, so arranged by Duchamp, came to represent an absent subject — a Buddha — in accordance with our general proposition. It would seem impossible to give an object the status of art and for it not to consequently invoke something other than itself. Thus, anything offered as art that is chosen for its intrinsic aesthetic qualities (such as a shark) seems immediately to acquire some additional representational or symbolic value (such as isolation, post-modernism, etc.) which it did not have in its own context. As we have demonstrated therefore, the practice of art can be generally understood as adding significance to matter by skilled arrangement. The 'quality' or value of the art subsequently depends on the quality or value of the significance added. The question now becomes: what does such significance consist of?

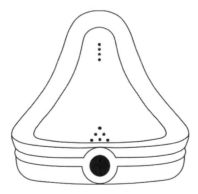

The significance of matter can be rearranged

The significance of art

The various words used to describe (primarily visual) artistic activity reveal something about its function. Words such as 'convey', 'express', 'draw', 'capture' and 'take' each imply a sense of transference or transportation.

- 'Convey' means to conduct, carry or transmit.
- 'Express' carries the sense of speed or expulsion.
- 'Draw' can mean to pull out, pull forward or take up.
- 'Capture' means to take prisoner, to remove, or transfer ownership.
- 'Take' means assume possession of (as in photography).

But what is being transferred between art objects and their human observers? The orthodox response would be meaning, although this would reveal relatively little about the processes underlying this complex cultural activity. In any case, the word meaning, like instinct, is often invoked where fuller explanation lies beyond our grasp. What then is conveyed, expressed, drawn, captured or taken in the process of making art?

Art as an energy regulating mechanism

The concept of art as a sophisticated system of data storage, or as a repository of meaning, provides only a partial account of its impact on the receptive viewer. A more inclusive explanation may be possible if we regard art as a transaction that conveys, expresses, draws, captures or takes energy. Proceeding from the claim that everything in the Universe is energy (cf. *The Post-Human Condition*) an 'energistic' interpretation of art allows us to be more precise about the kind of exchange involved.

(Figure 1)   (Figure 2)

(1) Michelangelo. 1501-04. David. Florence: Accademia (postcard).
(2) Cast of David, Victoria and Albert Museum, London (photograph)

These two photographs represent a statue of the young David (later King of Israel) by Michelangelo Buonarotti (1475-1564). Figure 1 is the original statue now resident in the Florence Accademia and figure 2 is a plaster cast on display in the Victoria and Albert Museum, London. Neither photograph conveys the colossal scale of the sculpture when seen in actuality (it stands about 4 meters above its plinth). Nor do they convey the electrifying impact of the fluent detail that describes every surface. Note that figure 2 (which is the view from which it was intended to be seen) emphasises the size of the lower hand and puts the somewhat large head into proportion. Although this work has been extensively discussed, Ernst Gombrich is one of the few to recognise what he calls the 'charge' that certain icons, images or carvings have. In the case of the *David* there is an erotic charge that is denied by the overwhelming attention given to the religious and civic significance of the piece. We recognise, however, that the erotic aspect of human nature is one of its most potent and there is little doubt that Michelangelo found the subject of an adolescent male nude erotically stimulating. This uncontroversial acknowledgement of the erotic energy embodied in the representation is but one of a number of possible 'energistic' resonations of the work. As an economy of energetic transfer between sculptor and viewer this piece conveys to us the virility of an active male about to slay the mighty Goliath by the force of will and slingshot, his athletic frame innervated with muscular vigour. The distortions, scale and resolution of its manufacture are traces on this Mystic Writing-Pad of an immense effort and concentration on the part of the artist, which can be detected by the viewer through the imaginative operation of mis-recognition and resemblance. The receptive viewer, sensitised by an appropriate cultural priming, would undoubtedly experience a thrilling, even 'electric', sensation in the presence of these absent men.

Art as an accumulator

Art objects are not only receptacles for, or containers of meaning, but also repositories of human energy. 'Inert' matter, such as marble, is arranged or modified through human intervention so that it acquires semantic value and 'captures' energy. This stimulates the imaginative resources of the viewer who interprets the representational meaning and the energetic charge invested in the work. The processes of mis-recognition and resemblance, essential to representation, conspire to excite energy in the viewer through the invocation of an absent object. In a work of art the effectiveness of the excitation of energy is proportional to the degree of energetic investment made by the artist in the material (lazy work is invariably poorly received). This might explain why intricate works often impress, although an artist like the late Henri Matisse, or a Zen painter like Sengai, might have argued that an even greater degree of concentration and effort is required to produce an image of fluent simplicity. That is certainly a Modernist belief.

Art and the unifying desire (1)

We have descried how the vocabulary of artistic practice implies transference of energetic states and how the extra significance added to matter can be read in terms of the investment and retrieval of energy. We have also discussed the factors of recognition and resemblance as they operate in representation and how, as with the cinema, our imaginations can be activated by sensory clues that allow us to perceive things that are absent. Taking these ideas together we may have come some way towards an understanding of the operation of art and its significance. Art, in common with all other technologies, removes obstacles to desire (often creating new desires in the process). The precise desires involved are many and various but certain ones predominate, in particular the compulsion to make and consume representations — a compulsion shared with photographers, filmmakers, writers and Palaeolithic cave painters.

Art and the unifying desire (2)

By representing something (in whatever medium) we invoke its presence in spite of its absence; this much has been clearly stated. By invoking something's presence we find, to our surprise and delight, that it actually becomes present — it arrives! Just as I cannot meaningfully use the word cat without invoking something of a cat, so I cannot lay down a line that resembles a head without conjuring the presence of a head, however crudely. Being a head, it shares perceptual characteristics with other heads (real or imagined). In other words, the material of representation takes on some characteristics, by imaginative association, of the thing it represents; paper and ink become a head, marble becomes a young man — at least in imaginative reception. Put more formally: The specific arrangement of energy that distinguishes a thing (x) is copied by a representation (y). Therefore, y is constituted by a similar arrangement of energy that distinguished x: — x and y are, in some sense, equivalent.

Art and the unifying desire (3)

To return to an earlier question; what then is transferred in the art object, what is conveyed, expressed, drawn, captured or taken? We might answer this by saying: some part of the thing being represented is transferred to the medium of representation (paint, ink, marble, magnetic tape, etc.). Since the consequences of a thing constitutes part of the thing (as we argued in Section 2) so the representation of a thing is consequent upon the thing and, therefore, part of it. Put another way, the depiction and its subject are continuous. This might explain why people who are afraid of spiders are even squeamish about touching rubber ones. The more potent the depiction the more continuity exists between the material of representation and the thing depicted — to the point where, in some cases, they become perceptually indistinguishable (as is supposed to be the case with certain wax-works). Of course, one effect of this is to extend the subject of the representation beyond the previous extent of its existence — one distributes things by representing them.

This is a pipe

Making representations involves organising matter so that it evokes something other than itself, simultaneously transforming it into the thing it evokes

Art and the unifying desire (4)

When we make, own or perceive a representation of a thing (a head or a body) we are dealing with part of that thing itself (inasmuch as any consequence of a thing is part of it). For example, the form of a statue of a male body is consequent upon the form of male bodies (or a particular one) and, thereby, continuous with them. In this we find a primary motive for the act of representation, in all its forms — the desire to experience something at will, to own it or to gain the experience of it, since by experiencing the depiction one is experiencing the thing itself (even if crudely or in part). By this movement the circuit of transfer is completed from the things being represented (male bodies), via the materials of representation (carved marble), to the recipients (viewer). In this sense, a direct line of energetic descent exists through the imaginative energy transferred from the subject by the artist to the material that is then retrieved the viewer (in apparent accordance with universal laws of thermodynamics).

Art and the unifying desire (5)

A depiction sustains our attention (albeit temporarily and unsatisfactorily) by uniting the depicted thing, the technology of representation and the perceptual apparatus of the viewer through the agency of the imagination. As with the operation of cinema, the technology of representation invites the viewer to become mentally and physically continuous with the specific energy of thing represented, thus experiencing something of the response that would normally accompany perception of the thing in reality. For example, the erotic arousal experienced when perceiving a nude male in the flesh is concurrent (although not identical) with the arousal experienced by the perception of a nude male in marble. The object of art is able to capture, store and transfer these energies by making itself into the object it represents and, thus, taking on the attributes (at least in part) of its subject. It is in this sense that ownership of the specific experience is attained by transfer of the significantly arranged material.

The art machine

We have given an account of art in terms of what is does rather than how it appears. One of its major functions is to act as a kind of consensual consciousness embodied in materials that are manipulated and arranged by skilled artisans. The process of arrangement inevitably infects the materials with some energetic resonance of the artist, and this can later be detected by a sensitised viewer. In this sense, art is technology through which we can deposit and withdraw aspects of our imaginative existence and desire: eroticism, holiness, civic pride, transcendence and humour. In addition, we have suggested how the energetic relay can effect a transfer of desire through the ownership of representations, and also how the material of the mark influences our perception of the value it embodies. Art becomes one of the many technologies by which the less tangible dimensions of human experience transcend the apparent physical limits of our bodies, just as candlelight extends our perceptual apparatus.

Drawings also record the trace of the energy used to make them

Section Seven

PROPOSITION:
The idea that livings things consist of organised energy is disavowed by conventional science.

Energy can be manifest as form

Modelling human being

There are a number of prominent models of human nature that are used to explain our experience. Those that try to reconcile high theory with the quotidian often produce partial accounts that overlook the complex dynamics of energy in all aspects of life. For example, biological deterministic models claim descent from Darwinism in positing 'genetic' or 'evolutionary' reasons for human actions. Sometimes called 'neo-Darwinists' or 'Social-biologists' such protagonists use their theory to interpret cultural activity in terms of how genetically determined behavioural traits affect the survival and viability of the species. At its most ludicrous and extreme it argues that acts of sexual violence are accounted for by the

107

male imperative to pass on their genes. Alternative theoretical models give much more weight to social and environmental factors, such as how ideological structures and economic imperatives determine human behaviour. Whatever the relative merits of these often mutually antagonistic models of human experience, they are both deficient in one respect: just as the energetic implications of art tend to be ignored, so explanations of human experience that recognise the significance of energy have been marginalised.

The controversy of the ' living spark'

In spite of neo-Darwinism, however, there is wider acknowledgement that living humans appear to possess a 'spark' or 'charge' — an essential resonant quality that distinguishes them from inert, dead or mechanical things. The most overworked depiction of this idea is Michelangelo's *Creation of Adam* in the Sistine Chapel, which persists in the popular imagination as a demonstration that life is fundamentally energetic. More generally, this idea of 'living energy' has been variously named and closely investigated over the course of human history in mythic, religious, scientific and artistic discourses. It is often described as a kind of 'heat' or emanation that is difficult to define or isolate, or as an invisible organising force that guides bio-chemical events. The presence of an urgent energetic motivation to life has occupied many notable intellectuals, for example, Henri Bergson postulated the élan vital in *L'Evolution Créatrice* in 1907 and Sigmund Freud's 'libido' was theorised as the instinctual sexual drive in various works from 1905 onwards.

The energetic debate

As we shall show, the debate about the existence and function of living energy provokes a degree of controversy. In this section we outline some of the resistance that an energetic conception of life meets from the authority of science.

The mechanisation of life

One of the defining characteristics of an organism, as it is now understood, is that it absorbs, consumes and expels energy through various bio-chemical processes. Given this, the presentation of life as organised energy has been underplayed by recent prominent theorists, particularly by neo-Darwinists such as Steven Levy and Richard Dawkins. Levy and Dawkins are mentioned, not because they represent the most advanced state biological ideas, but because they represent a certain dominant orthodoxy about the status of current scientific thinking. These neo-Darwinists favour a model where life is seen as organised data, or information. This is almost certainly because an informational paradigm chimes with the contemporary enthusiasm for digital information processing, and for which the DNA model of life provides a literal justification. Looked at from a postdigital perspective this informational model can seem restrictive in that it tends toward the notion of life as a machine displaying only mechanical behaviour (even with the proviso that such behaviour may be very complex). This can lead to the error of seeing other emergent phenomena in such informational terms, a tendency that is common in brain studies. From this position it is quite easy to see how informational models of living processes can become metonymically confused with actual processes — the organic whole is displaced by the abstracted representation.

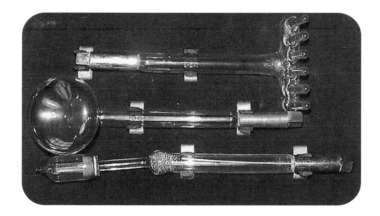

Electrical therapy technology

A note on the living 'machine'

The mechanical trope that depends on a direct correlation between cause and effect to explain things is regarded as incompatible with an energistic world-view (indeed, mechanists insist on nothing less than its subordination). Steven Levy dismisses the vitalist 'entelechy' thesis proposed by Hans Driesch in the 1890s: "Though it would be a difficult task to unearth a modern scientist subscribing to that true doctrine, vitalism of a sort seems to persist. There is a peculiar reluctance to concede the honor of life-form to anything created synthetically." (Levy) We see here again, in Levy, the tendency to conflate the synthetic with the real. The synthetic, which is currently mechanical, is then transfigured into the real such that the real itself becomes mechanical. The preferred notion of life as a self-replicating machine proposed by mechanists such as Richard Dawkins and Steven Levy sounds suspiciously like a hangover from the early Industrial Age when the clockwork metaphor of reality flourished. One reason for the persistence this metaphor was surely the excellent rate of return on capital that machines delivered and upon which all the industrialised economies prospered. Today, in the information market, the computer metaphor flourishes and no doubt protein tropes will take hold tomorrow as transactions become bio-degradeable and genes define wealth and health.

Postdigital life

It would be simplistic to suggest that Dawkins and Levy are crude mechanists who deny absolutely the complexity of life. Indeed, they subscribe to fascinating and plausible notions of complexity, which make for seductive copy. They variously describe how the appearance of living or thinking matter can spontaneously emerge from conditions of sufficient complexity. Dawkins' *biomorphs*, Levy's *artificial life* and Daniel Dennett's *conscious robots* are potent demonstrations of the idea that processes previously regarded as magical, or beyond explanation, can be convincingly modelled in non-organic substrates such as a digital computer. That is, they can be modelled in a machine. Simply put, they suggest that if life and consciousness can be modelled on machines, then they must, categorically, be machines. From a postdigital point of view such a conception of life and consciousness is problematic, if only because the metaphor of the machine has become so 'universalised', describing processes ranging from the laundry to the nano-sphere, as to render it almost meaningless as a valid category.

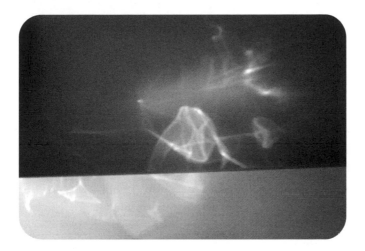

∧ digital photograph of an apparition

The mechanists and the mystics

In broad scientific terms the possibility of 'living energy' remains a speculative theory in the absence of solid evidence to establish it beyond doubt. However as we have suggested, in some circles the notion of living energy is summarily dismissed as a regressive myth since it contradicts the whole thrust of contemporary science. For scientists like Steven Levy the idea of living energy reasserts the phantom of vitalism that science was supposed to have laid to rest. In considering this meta-discursive aspect of the question we are drawn into the vortex of an old controversy, broadly described as that between the 'mystics' and the 'mechanists'. Daniel Dennett, for example, in *Consciousness Explained* (a misleading title if ever there was one) equates notions of *deus ex machina* and *élan vital* with 'magic' and 'mystery' and contrasts them unfavourably with the morally superior 'reproductive machinery of DNA'.

The reliance on science

There was a basic antagonism throughout the twentieth century between powerful institutional authorities of rational science and the unorthodox beliefs of the vitalists, mystics and homeopaths. Yet the assumptions underlying this antagonism are now being rapidly revised as the authority of science (by which we mean the accumulation of institutions, bodies and authorities that define themselves and each other as scientific) is increasingly questioned. Given that some prominent scientists claim their discipline to be the path to universal knowledge, such totalising tendencies start to look arrogant and over-optimistic. They do not allow for the possibility that huge strata of reality may be impervious to scientific analysis. Should we accept, then, that a phenomenon has no validity unless it has been legitimised by science? This would be absurd since any reasonable scientist would readily admit to the limitations of the scientific method (there is no scientific explanation of 'whimsy'). Although science is only a convention for investigation, it has progressively become a benchmark of truth — a truth apparently verified by logical argument (see note 6). The co-operation of science and logic over recent history has tended to reinforce the idea that the two are inseparable. For some philosophers of science the validity of a scientific idea demands a logical corollary (Albert Einstein, for example, was never able to accept the essentially illogical explanation of the quantum mechanics offered by the Copenhagen interpretation). This view inevitably stems from a belief in the essentially logical operation of all reality, including human behaviour.

The discourse of logic

Logic might be a human construction but this does not mean that humans are logical. As has already been pointed out, logic is a subset of our linguistic protocols — a self-referential idealisation imposed upon reality. If it were the only criterion by which we measured the validity of ideas we would be infinitely poorer. There are many instances where the application of reason has led away from valid solutions rather than toward them. Karl Marx and his followers purported to be engaged in the scientific analysis of society and economy. They thought they had uncovered a set of underlying laws that drove social organisation inevitably towards the

112

eradication of internal contradictions and the dictatorship of the proletariat. By a process of extensive research, brilliant analysis and logical argument they arrived at some conclusions about the progress of industrial society that were used to validate of some of the most inhuman regimes ever to exist on Earth.

Let us take it that:

a. Phenomena do not necessarily have to be logically explicable to be true (useful). Quantum mechanical events are regarded as being, in many instances, logically incoherent, yet they are taken seriously.

b. Phenomena may still have a valid existence even if they cannot be empirically tested or substantiated. Our lack of ability to see what is invisible, or to know what is unknown, does not preclude the existence of either.

Given these two entirely reasonable propositions, we are able to speculate with a certain amount of legitimacy about the phenomenon of a living energy. We can do this despite there being no complete materialistic theory of what this energy is, and despite some prominent researchers actively dismissing it as tantamount to the evocation of magic.

The 'new age' of ideas

Despite the ongoing success of scientific materialism in validating the logic of industrial growth (and stoking its flames), some of its more extreme apologists who see life as programmed matter, and the mind as a digital machine are starting to sound a bit quaint, even embarrassing, in their zeal. A subtler and more expansive intelligence is beginning to emerge from the wider synthesis of ideas made possible through new communications. Whereas 50 years ago scientific intelligence was restricted to a closed circuit of journals, institutions and conferences (many sponsored by governmental or commercial interests) it is now more widely circulated through press, books, the internet and TV. This increasing promiscuity of ideas and data inevitably leads to cross-fertilisation (some might say a 'bastardisation') between diverse branches of knowledge. One offspring is the rapid growth in 'future sciences' pointed to by the various conglomerations of theoretical physics, nanotechnology, cryogenics, genetic modification, psychic research, cosmic spirituality, telematics, extra-terrestrial speculation, environmentalism, artificial life and intelligence, healing arts, and so on. Myriad examples of such tendencies are found on the web and in newsgroups. It is apparent that very few of these integrated ideas allow for the possibility of hard materialism of the kind expounded for much of the late nineteenth and twentieth centuries. It is impossible to predict where this fusion of ideas and approaches might lead, but it is likely that it will lead away from where we are now.

All things resonate

It is not necessary to subscribe to pseudo-science, or New-Age nostrums to appreciate that everything has energetic resonance. Orthodox contemporary theories of physical reality describe matter at the smallest detectable levels as strings or packets of energy. In these theories, solid things are made up from huge matrices of tiny vibrations, leading to the widely accepted view that matter is organised energy. Add to this the wealth of resonant energy that we can perceive at the human level, in the form of heat, light, sound, gravity and radiation, and one can appreciate that all things, animate or inanimate, are compounds of energy.

Energy as leaf

Energetic tropes

Even materialists who dismiss vitalism often recognise that living things are energetic in the language they use, just as genetic determinists subscribe to the idea of energetic conservation to account for behavioural traits. We betray our belief and investment in energy in a number of areas of everyday speech: we speak of individuals as 'brilliant', 'bright', 'exhausted', 'glowing', 'explosive', 'exhausted', we have metaphysical concepts like 'élan vital', 'the ghost in the machine' and 'sparks of inspiration' (seeing the light). In relation to large-scale systems we talk of 'dynamics', 'meltdown', 'overheating' and 'cooling off'.

Underground energy

More formally, the existence of a 'living energy' is foundational to the doctrines of several ancient traditions of religious and scientific knowledge (Taoism, Buddhism, Shamanism, Voodoo, indigenous American and Australian cultures as well as others). The tendency of Western materialist rationality to dismiss these traditions as 'primitive', 'superstitious', 'unprovable' or 'mystical' betrays more than arrogant ignorance. It underscores the implicit anxiety felt towards unmediated forms of energy which was manifest in the 'racism' that scientifically classified 'Negroes' as 'natural slaves' in the eighteenth century. As we have already suggested, those with authority over the manipulation and distribution of 'knowledge' are generally squeamish about the role of 'spirit', 'soul' or 'vitalism' in the operation of life.

Universal energy

The postdigital offers an energetic model of living processes, rather than programmed information, a model fore-shadowed by some older, more inclusive visions of human experience that we have previously enjoyed. In the nineteenth century, for example, electricity occupied a special position in respect of nature in that it was seen as significant for both the Spiritual and the Scientific. Other forms of energy (heat, light, sound and gravity) had always been commonly experienced, whereas electricity had a special status in that did not seem to reveal itself naturally, but required induction or invocation by specialised apparatus. It was this mystery, and the belief that electricity was a manifestation of some divine life force, that induced the sick to invest in medical treatments linked with electrical charges. The connection of electricity to organic systems, particularly the human body was not entirely the product of ignorance and superstition. At the close of the nineteenth century electricity was regarded as a uniquely self-reflexive power, which had special position because of its intimate connection with almost every other form of natural energy. In addition, electricity was a technology which could 'write itself', so to speak, in that it seemed capable of reproducing the conditions necessary for its own production. Public demonstrations of dead limbs twitching under electrical stimulation had long been a scientific entertainment, which seemed to confirm the idea that Michelangelo had depicted several centuries earlier in *The Creation* — the equivalence of energy, life and creativity.

Religious illumination

Cultural notions of living energy

The idea of a living energy, manifested as libidinal, sexual or creative urges, is so widespread in human culture that some people have argued that the very foundations of social organisation are based on its systematic repression. Others have claimed that its potency, and potential to destabilise, is managed through creative and intellectual activity. Whatever the merits of these claims, behind them lies a tacit recognition that dynamic forces shape humans, their cultural artefacts and institutions. Indeed, it would be a fruitful project to consider an energetic history of human culture (see note 7). In some (slightly unorthodox) scientific circles today the generic term for such energy is Human Energy Field (HEF) which, although not without its share of sceptical criticism, is an active branch of investigation in many countries. HEF can be seen to have had its precursor in the ancient Yogic concept of *prana*, which is recorded in texts about 5000 years old, in Chinese writings about *Chi* which have been in existence for over 3000 years and the Jewish Kabbalistic notion of astral light, which can be traced back to around 600 BC. Later, the idea of living energy being a form of heat or light is represented as the glow, or halo, emitted by sacred icons in Christian art as well as Buddhist and Hindu imagery. It is not necessary to subscribe to these systems of belief to appreciate the influence that an energetic conception of life has exerted on such civilisations.

Institutional resistance

Given the deep cultural prevalence of the energetic idea, it might be surprising that it is so readily dismissed by many conventional scientists. Although prevalence of belief is not necessarily evidence for something's existence, nevertheless the historical and cultural persistence of these ideas, even when antagonised by contemporary science, is remarkable. In our rationally enlightened times house purchasers are increasingly likely to require an 'energy' survey from a *Feng Shui* consultant as well as, or even instead of, a structural survey before buying a property. Professor Graham Thompson a particle physicist from Queen Mary's College, London claims that "There is no scientific evidence for Feng Shui. I have studied energy for years and there is no such thing as Chi — the so-called life-force upon which Feng Shui is based" (*Daily Telegraph*, March 4th 2000). Despite such authoritative dismissal of Feng Shui, the growth of related practices like massage, colour and acupuncture therapy, crystal healing, dowsing and yoga — all activities that postulate a flow of energy in and around life — continue to attract the patronage of rational and prudent consumers. Whilst easily dismissed as middle-class fads, that would hardly explain the implicit rejection of scientific authority that such beliefs represent.

Apparent energies

Strong attractors

For some observers it seems that the fashionable alternatives to, say, conventional medicine provide evidence of an increasing resistance to scientific ideology, or at least the claim of science to be acting neutrally. The ideology of science is seen to be represented by conglomerates of mutual self-interest — medical colleges and foundations, government institutions, drug companies and those in positions of scientific authority who reject the validity of unsanctioned practices. Collectively, they may determine that certain activities and theories remain unrecognised for a number of self-serving reasons. While many, if not all, individual scientists are nobly 'searching for truth' their work is largely funded by, and used to support, institutions with less claim to neutrality such as medical foundations, highly capitalised drug companies, academic departments, unregulated administrations and corrupt governments. Whatever the reasons, consequent public cynicism about science was highlighted in a report published by the British House of Lords in

2000. It spoke of a general disenchantment with scientific authority on the part of 'the public', i.e. those whose taxes and expenditure fund most research. Such reports are evidence of a growing anxiety about the decline of scientific authority and what might replace it. This lack of confidence in many aspects of conventional science may account for why alternative explanations about the world are increasingly embraced. Yet the differences between opposing world-views may not be as clear as some would like to think. Practices like Feng Shui and conventional science require little legitimacy from outside of their own terms of reference and both are understood as rigorous disciplines that use empirical methodologies to describe and modify the world. Given these fundamental similarities, it seems odd that they should be regarded as so incompatible.

What is true is what we believe to be true

The proposition that living forms contain, or emit, an energetic field is supported more by weight of tradition than it is by scientific evidence. Which is most detrimental to its legitimacy: the wealth of legend and folk-lore or the lack of scientific theory or evidence? Are we not inclined to treat with greater scepticism that which attracts the most 'irrational' belief and superstition? Whether or not it eventually transpires that the existence of a living energy can be scientifically determined is largely immaterial. What is more significant is the extent to which people believe it to be true and the more empirical or anecdotal evidence that accumulates in its favour the more this will be so. It is not our intention to prove either way the truthfulness of any claim made. What is more useful is to understand that human belief rarely arises without provocation and that superstitions are almost always a social response to perceived reality. As with myths, they often betray the operation of human desires and fears that some may regard as irrational, and others would call intuitive. Take for example the report that Police chiefs in the city of Nottingham (UK) initiated a "crime prayer-vention scheme" [sic] in which local congregations were asked to conduct 'targeted prayer' for the lowering of crime rates on problem housing estates. According to the Police, so successful were the results in reducing the crime levels, with measured falls of 10%, that the scheme was being extended to other local estates (*Guardian*, March 23rd 2000). The report ends with arch-sceptic Richard Dawkins (presumably in his appointed role as Professor for the Public Understanding of Science) commenting that, "If the Police of Nottingham are as gullible as this, it's a wonder they ever solve any crimes at all." One could argue that Dawkins' attitude is profoundly anti-scientific, since one should at least retain an open mind pending further investigation rather than dismissing the evidence *a priori*.

The errors of belief: the intervention of the observer

The consequences of believing something false to be true, or something true to be true, often amount to the same thing. The external verifiability of a claim can be irrelevant to whether or not people believe it. Even when something is shown to be false, people will often still think it true. Most historians agree that, inasmuch as Jesus of Nazareth existed, he was certainly not born on December 25th in the year zero. The reasons for believing in something may rest on more than just mundane fact. Psychologists refer to the phenomenon of 'suggestion' to account for, what they would regard as, the frequent confusion between belief and reality and what others might regard as a deeper connection between the two. In a classic book on unorthodox beliefs and psychic fraudsters, Joseph Jastrow outlines the history of the Ouija board. The original device, described in Roman accounts, was a pendulum of metal and string held over a board of letters. He recounts how, subsequently, many investigators noted an uncanny phenomenon in which the pendulum would move when held over certain substances and in differing directions according to which substance it was held above. It became something of a curiosity in physics and attracted the attention of the French chemist, Chevreuil, in the mid-nineteenth century. His investigation initially confirmed what had been claimed. When he held a pendulum over a plate of mercury it started to swing, and stopped when a glass plate was interposed between the pendulum and the mercury. He tried to eliminate a number of possible causes, using different substances and steadier supports for the pendulum, but still it moved when held over substances in contradiction of the then known laws of physics. Finally, he blindfolded himself, then had an assistant insert the substance beneath the pendulum and then recorded the result. To his apparent delight, the pendulum stayed still. Jastrow regards this as the first recorded instance of, what would later be called in Psychology, the power of 'suggestion':

Chevreuil had made a great discovery which stood by him for life. "So long as I believed the movement possible it took place; but after discovering the cause I could not reproduce it." He adds that the experiments "might be of some interest for psychology and even for the history of science". They show how easy it is "to mistake for realities, whenever we are confronted by phenomena in which the human sense-organs are involved under conditions imperfectly analysed."

However, this conclusion in itself suffers from a kind of implicit suggestion. We are meant to assume the given interpretation explains the strange movement of the pendulum, but this triumph of the materialist mind over mystery does not really explain anything. The 'Placebo Effect' is well known to medical researchers. Some patients suffering symptoms are told they are being treated with a certain drug and will respond to it even if they are never actually administered the drug at all. Yet although this effect is widely accepted, it does not explain why people respond to non-existent drugs simply because of the belief that they have been taken. It simply demonstrates a link between belief and consequence, just as Chevreuil's experiment does. The explanation for why the pendulum moves, or for why the absent drug works, remains a mystery.

Mélodie d'Ouija

Science as energy

The inconsistent and restrictive prescriptions about science that certain leading scientists assert are supported by recently revised histories that masquerade as authoritative. For the better part of the nineteenth century the dominance of professional science in determining truth was not absolute, nor did its progressive claim to authority go unchallenged. Indeed it was not until the twentieth century that the supremacy of science became the orthodoxy. In 1939 Alexandre Koyré argued that the history of science was essentially a history of ideas, and this prescription came to underwrite the modern belief in science. His view was essentially a revision of earlier descriptions of scientific practice which were much more contingent on everyday experience. Koyré's history of ideas tended to ignore the evidence of other (more physical) methods of research, such as laboratory practice and working notebooks, and focused instead on professional historical interest in scientific conclusions. In contrast to Koyré's hierarchical methodology of proceeding from abstractions (ideas) to outcomes (inventions), more scholarship has examined the history of experimentation that formerly had been neglected on the grounds that it was of secondary significance. Stephen Gooding, Trevor Pinch, and Simon Schaffer, among others, have approached science as both a mental act and a physical process. They trace its practice as combined thoughts and actions by looking, especially, at the evidence of experiments and laboratory notes. This emphasis on knowledge as action has directed attention away from well-honed histories written from the vantage-point of scientific outcomes. These tended to overlook the patterns of actions, contingencies and responses to failure that ultimately contribute meaning to an experimental practice. Such 'new histories' of science may be criticised for their over-valuation of experiment and their somewhat speculative claims about mental processes. However, they do attempt to marshal hard evidence from notebooks to show how the processes of scientific practice shape the final conclusions. More than anything they demonstrate that science, like art, is a process of transforming energy between existing states in order to satisfy human desire.

Science in the postdigital

Science is now understood as a specific set of ideas practised through an arbitrary set of methodologies which has privileged a materialist view of the world at the expense of an energetic conception of existence. Although the scientific method has, in many respects, enriched and extended our lives it has also, by necessity, diminished the experience of our own existence. The desire to render the chaos of experience as both stable and reproducible, independently verifiable and immune to the distortions of perception, was noted in respect of medieval Italian painting. It emerges again in the scientific method and, by extension, in philosophical descriptions of the world. Thus our appreciation of the world and the phenomena it presents us with is necessarily diminished. Electricity, for example, which emerged as a manifestation of a universal living force is transformed into a crude commercial product that drives forward subsequent stages of industrialisation.

The postdigital model

The purpose of this section is not only to argue there is current lack of theoretical emphasis on life as organised energy, and that it may be valid to challenge the authority of science, but also to negate the binary opposition between 'mystical' and 'material' stances on observed phenomena. Whether we consider the strange motion of a pendulum, or the operation of life itself, the postdigital conception points to the need for a new, subtler, model of understanding than is currently offered by those in intellectual authority.

Recorded energy

Section Eight

PROPOSITION:
A recording is part of what it records

What happens when we refuse to separate the consequences of things from their material or conceptual form?

Any consequence attributable to a thing is a constituent of it

One of the chief questions posed in this book has been what happens when we refuse to separate the consequences of things from their material or conceptual form? As we pointed out earlier, the light, smell, noise and heat emitted from a candle is much the candle as the wick or the wax. The candle and the consequences of its use (an illuminated lover) become indistinguishable - the lover becomes part of the collective phenomena that make the candle what it is. This idea refutes the notion of separate individual things that work upon each other by cause and effect. Instead, by regarding any consequence attributable to a thing as a constituent of it we challenge the accustomed view of reality. By this understanding, the observer is clearly continuous with that which is being observed.

Overcoming the observer with a representation

Previously we discussed how Alan Turing claimed it was theoretically possible to build a Universal Machine that could produce its own likeness in every respect, thereby obviating the need for human validation. This was more than a technical challenge since it undermined a cornerstone of modern science. The insistence on the singular status of the human observer in science demands a distance between the object and the subject for truth to be verified. The interference of the observation process must be minimised and, for this reason, experiment became the preferred method of testing. Only with this distance could a scientific claim be shown to be true regardless of the observer's position. This was a state of affairs that the Universal Machine challenged since it was both the object of the claims and the subject that verified it.

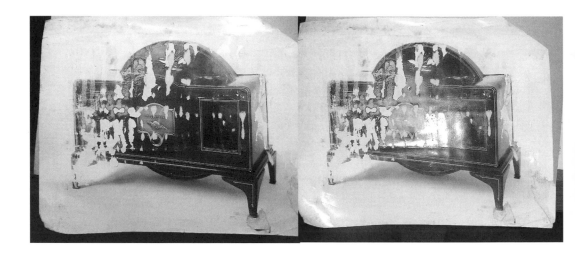

Digital photographs of a picture of a 'Televisor'

Observers and experiment

In the seventeenth century the social standing of the subject was thought to confer some guarantee of objectivity. The vulgar man was not credited with knowing himself sufficiently to be impartial. Knowing himself seems to have meant being able to separate the mind from the body, and know when the body — regarded as an interference in the intellectual life of men since Plato — may be distorting the observations of the world. This became unsustainable as it was shown that, whatever their social standing and however highly educated, the expectations of the experimenters predisposed them to interpret the results in a way that favoured their suppositions. Without any intention to defraud it was noticed, time and time again, that the observer of an experiment could exert influence on its outcome.

Descartes

Modern science did not regard this power of suggestion over the material world as a positive force to be reckoned with. Instead it chose to regard it as a shortcoming in the perceptual apparatus of the observer, a delusion to be eliminated. If the senses could be fooled and were untrustworthy, as Chevreuil had declared, then only a suspicion of the senses could affirm the objectivity of the experiment. At the root of this belief was, of course, the philosophical basis of modernity, in which the body was of a different order to the mind in that the latter could observe and affirm the former, but not vice versa — quite unlike the Universal Machine.

Comparing images

In the fifteenth and sixteenth centuries, attempts to sustain an objective view of the Universe were made though the design of scientific experiments and apparatus, some of which were subsequently used by artists (many of whom were also scientifically curious). Prominent amongst these devices was the *camera obscura*, which guaranteed the authority of the draughtsman's image to a level of scientific accuracy. As a scientific tool this might have led to ways of representing reality independently of the observer (devoid of perceptual distortion or extraneous meaning). In the hands of artists, who where appealing to a different constituency,

127

the apparatus allowed the mundane to be invested with extra significance. What was clearly emerging was a belief in the authenticity of images in comparison to the visible world.

Depictions become real when they are recognised as something

Making pictures of things

During the period of Western art in which painting was most prominent, artists mostly preferred subjects with some personal significance, and it was often said that choice of subject was 'inspired'. Artistic inspiration arose from the desire to 'acquire' or 'capture' the thing the artist saw or felt. Yet what was metaphorically acquired in this transference of ownership? What was being captured or transacted in order for the painting to become continuous with its subject? In Section 6 we argued that what was being 'taken' or 'drawn' was a specific arrangement of energy.

> Writerly painting. When portraying important objects, one will do best to take the colours for the painting from the object itself, as would a chemist, and then to use them as would an artist, allowing the design to develop out of the distinctions and blendings of the colours. In this way, the painting acquires something of the thrilling innate quality that makes the object itself significant.
>
> Friedrich Nietzsche, *Human, All Too Human*

Capturing ourselves

A particular arrangement of light is unique to any object of inspiration (what is being captured) and this constitutes, to some extent, the object in itself. Today, when we capture our own photographic likeness we expect the camera to collect a particular arrangement of light, shade and colour in order to make a convincing representation of our face. This process produces a unique image that, both metonymically and substantially, represents our face: the image both stands in for us and is part of us.

What I am

Aside from all the 'dog-eared' debates about representation and meaning, one implication of energetic specificity is that the arrangement of light caught by the camera from a face is, in a phenomenological sense, a constituent of the face. In fact it could not represent (in both senses of the word) unless it was part of what is represented since all things are continuous with their representations. Something of a person can be said to exist in every image they appear in, just as something of a person is deposited in every other consequence of their existence. We are more than the weight of bone, tissue and fat by which we might be defined by medical science; we are also the sum of all events arising from our existence.

What is it of me?

The picture of the world that is informed by this extended view of human existence is, at once, beautifully simple and fiendishly complicated. It is simple because everything becomes one thing and we no longer have to conceive of the world in dissected, fragmentary terms. It is complicated because we have to accept that our conception of things must include their infinite consequences. Consequently the materialised subject, as it appears say in a photographic portrait, is more than a pattern of stabilised photons. Above and below the reflecting layers of skin are the resonating thoughts, feelings and experiences that a person feels and that their face expresses. In other words, photography does more than change halides from one state to another in response to varying intensity of light. The camera (with its

129

peripheral apparatus) is not an inert machine but an energetic conductor fully implicated in the emotions it arouses. In collaboration with human intelligence, the camera does much more than merely secure a particular arrangement of light — it extends things, including emotions, beyond their original time and space.

Putting you in the picture

Written into the photographic portrait is as much of the subject as the medium will permit, perhaps more than a strictly mechanical view of the apparatus might suggest. States of being are evident in pictures when we say, "She looks happy in that picture" or "He looks sexy in that one". We recognise something other than simple identification from the pattern of light, colour and shade. An apparently 'inert' arrangement of chemicals reveals, and can induce, an immaterial state of being or a condition of existence. In this involuntary act of self-recognition, we are implicated in the circuit of energy that photographs induce and conduct.

The use of youthful, pretty, happy faces in advertising imagery has a purpose. (Advertising image from a toothpaste advertisement, LOOK magazine, New York, 1948, vol. 12 No. 20.)

Only with a perverse (or highly critical) reading could we infer anything other than a state of unified pleasure and health from the beings in the photograph above. This condition is conveyed to us because we mis-recognise the printer's plate, ink and the half-tone screen for contentment itself. From 1948 to whenever and wherever we are now, something immaterial has been conveyed, intact, by this American toothpaste advertisement, distorted only slightly by cultural interpretation. If the photographer wanted to secure a quality as ephemeral as disembodied health (with minimal ambiguity) could it have been done more directly than by recording light from apparently healthy dogs and people in an artificially arranged scene? This image displays faith (and a considerable amount of invested resources) in the idea that the ethereal can be stored and transmitted through the material.

Does a representation appropriate the essence of the represented?

Early portrait photography, perhaps more clearly than contemporary photography, demonstrates the directness of the energetic transfer between subject, apparatus and viewer. Although often rigid, grim and stony, early portraits can have a ghostly air of reality that suddenly collapses the distance of time passed. Something of this strange intensity may be accounted for by the slow emulsions which insisted on the sitter remaining motionless for up to an hour as the chemicals on the plate changed. Often it was necessary to clamp the head discreetly in order to restrain natural movement. These technical facts are frequently used to explain why there are very few images of smiles — that moment of spontaneous transformation of energy from mind to body — during this period (conversely it may also account for the insistence on smiles as emulsion became more sensitive). Of course the pretensions of middle-brow culture which inflected the uses of early photography played their part by emulating the postures of high art (Jesus must have smiled, but we cannot recall a painting that depicts it). Nevertheless, the fact that early photographic portraits were exposed over long periods contributes to a different sense of presence than one is used to from 'instant' photographs. The prolonged exposure seems to have allowed something extra to seep into the plate: by sheer concentrated effort a deeper trace of the slow-motion cathexis between the subject and the apparatus is visible.

131

After Paul Nadar, Portrait of the Chemist Chevreuil, 1887

Writing energy in and out

The production of an image, such as that used in the toothpaste advertisement, may require significant energetic investment that is not immediately apparent within the frame. Beyond the effort embodied in the mechanical design and manufacture of all the necessary photographic equipment is also the collective creative expertise that adds another dimension to the energetic economy. The skill and effort of the artists, photographers (lighting crew, technicians, et al.) or the effort of the pioneer photographer is also 'drawn' into in the visible transaction between author and viewer. The 'hard write — easy read' inauguration of such an image conceals the infinite complexity of the energetic economy as it is stored and released in fast and slow burns. The picture becomes a membrane in which the active exchange between the material substrate and the intelligence of the viewer emerges as 'meaning'. The photographic image absorbs and reflects the various energetic processes necessary for its realisation and thus makes that energy available to the viewer in overt and covert forms. The toothpaste portrait demonstrates how technology can distribute specific energy by trapping experience in a substrate that can be recouped through negotiation with the human nervous system and its imaginative apparatus.

132

Existing beyond

One especially striking, and accessible, example of the distribution of specific energy across time and space is that of the famous person. As with any representation, particular formations of energy are trapped in various media and then recouped through the imagination. As the media are distributed so are the specific energies trapped in them. In effect, the subjects being represented in the distributed media are themselves being distributed beyond their own time and location. The function of recording media is not only to trap patterns of energy but also to make them stable, portable and exchangeable. Fame (in the contemporary sense) is what results from the wide distribution of particular energy formations (people, places, objects). The effect of fame, in the case of a person, is to extend their presence (even in their absence) insofar as part of their specific energy is deposited in any representation made of them. Such people can acquire a certain immortality as long as any perceptible consequence of their existence remains. Whilst this is true for us all, it is usually more visible in those who have obtained wider recognition.

Some identities become commodities

The irony of fame

In practice most people have little say over how their own image is used. Exceptionally some identities are registered as a trademark (as was recently done by the Diana Princess of Wales Memorial Fund). This, and other attempts by famous footballers and film-stars to retain exploitation rights over their own image points to a complexity in the assumptions which underlie such attempts, as well as being further recognition of the continuity between objects and their representations. It is ironic, not to say paradoxical, that famous people often rely on the free and wide circulation of their image to achieve stardom. Once having achieved fame, they (or their managers) often attempt to manage the exploitation of themselves by registering their image as a trade-mark. If the continuity between objects and their energetic consequences is ignored by scientists, at least it is acknowledged in the law

A detour on Diana: the extended flesh

Princess Diana's death was a startling and perplexing mixture of sentiment, irrationality, and contradiction that underlines the ease with which the image encodes the actual. Although the scale of the public expression was unusual, perhaps even unique, some of its features were not unfamiliar. The least welcome resonance of such collective behaviour was with recent political history. The hysterical glorification of a Wagnerian hero/saviour figure used to rally the crowd inevitably raised questions about how different Diana's following was from other 'charismatic leaders' in recent history. She was politicised from the start as the modern face of the Royal Family, and after her divorce she had been used by various factions as a rallying point for a new republican movement — particularly in Australia.

For some sections of the press, intent on republican reform, she was the perfect agent for the cause. She was the wronged woman badly treated by the most privileged class who would never be able to replace the Monarchy she was being used to dislodge because the public staging of her amorous affairs eliminated her from the possibility of ever becoming Queen — at least as we currently know the function. Her romances had involved some rather sordid telephone stalking of married men, and her lovers had included stereotypes of bounders guaranteed to raise the public hackles: her bodyguard (Barry Mannakee, a mere Sergeant) a gormless cad of an army captain, an English rugby player who was handsome but dumb (allegedly), and to cap it all for racist Britain, a Muslim. Diana the deposer could be discredited at any moment convenient to those manipulating her image as a loose woman with a penchant for rough-trade. It was this vulnerability which gave a woman who had political influence and considerably more, emotionally and materially, than any individual might expect out of life, the air of being a victim. On top of this, something about 'the big lie' and the massive following by people who were apparently oblivious to the manipulations was vaguely sinister for Europeans with an eye on recent history. In her life she was a vulnerable and dangerous victim with a massive material advantage.

The uncomfortable parallel between the crowds in a public park mourning a powerful person claiming to be a victim and other totalitarian cult hero worship of the past did not pass without comment. The anxieties were dispelled, however, by the theatrical excess of the sentiment over a person who was clearly subject to widespread and variable interpretation and lacked gravitas. What was for some people a glamorous heterosexual woman's role model was for others a gay icon of self definition; for others she was intelligent and cultured but also someone who like pop music and was at home with the stars from *Hello*! magazine. Yet others saw her as caring human being only with thoughts for others despite her evident self obsession as she succumbed to eating disorders, worked out in Chelsea and subjected herself to all kinds of quack treatments. From the popular newspaper archive alone as much evidence could be staked against any of these claims ('real' photographs, fashion victim, a selfish shopper, health-freak, etc.) as could be marshalled to support them. She was too fragmented to be a new Rienzie but such inconsistencies and contradictions did not lead to incoherence; the very gaps in the Diana story it seems gave rise to a following devoted to weaving narratives around personal detail however irrelevant or prurient. She was both fashion plate and pin-up — an image waiting for a text, a star shaped by individuals and the press less for her own qualities than the needs of those who modelled her. This kind of interest precipitated the sort of fan-fiction usually reserved for soap operas which interprets and reinterprets every shred of evidence to produce a coherent whole that is quite independent from the original (indeed some of slash fanfiction wilfully distorts the original material by insisting on cross dressing the characters). In her death the contradictions which constituted the public Diana (actually no more than a consequence of the way data her about her was organised) became the perfect extension for the energetic imagination of conspiracy theorists and a necrophiliac investment of human energy in dead affairs.

The Bell-Hop's tears keep flowing

Fame of the kind experienced by Diana depends on intensive investment in the production and distribution of representations of the person. As these are consumed, (often involuntarily), they form a depository of congealed ideas in our imaginations. The person is identified by these ideas and is continuous with them. For example, we may have our own compound matrix of images and thoughts about Elvis Presley that enable us to immediately recognise representations of him, all of which are consequent upon his existence. This personal state of consciousness, however, is ultimately managed by Elvis Presley Enterprises, Inc. in as much as licenses are required for commercial, and some non-commercial, use of his face, name and signature, whereupon a royalty is payable. Nearly all representations of Elvis that we might encounter are subject to legal and financial control, even if we are not directly implicated in the commercial transaction. In effect, there is a price on his head despite the fact that his image only has value by virtue of the qualities invested in him by us in the first place, which is then sustained by our collective imaginations and desires. What does the Elvis fan get when they purchase a cushion from the gift shop at Gracelands in Memphis, Tennessee? As a souvenir (in French, a cheap memory) it can be no more than a reflection of our own mind. In this kind of fan culture, we seem to have lost few of the tendencies that fuelled the international trade in holy relics. The significance of otherwise insignificant objects was conferred by virtue of their continuity with the venerated figure.

The cultural product as imaginary membrane

The cultural product (the printed cushion, the photograph, the portrait, the film, the diagram or the recorded song) is a kind of membrane of the imagination that merges two sets of circumstances. On one side are all the events that are being portrayed (the people, light, clothing, scenery, objects, and so on). On the other side are all the events that make up the use, or reception, of the product (its means of distribution, viewing context, condition of the viewer/listener, and so on). Between them lies the paraphernalia of representation — the energy transfer devices (painting materials, cameras, recording equipment, processes of storage, replication and retrieval, and so on). These devices operate within, and generate, the space that connects and separates all these circumstances. The membrane connects by virtue of its commonality to both sets of circumstances, which could not otherwise be brought together in such a reliable way. For example, the scene recorded by the camera in the 1948 toothpaste advertisement becomes connected to our experience in the present as we look at it. But the membrane also separates both sets of circumstances. The events 'before image' and 'after image' remain distinct. We cannot fully recover those 'pre-photographed' occurrences in the 'post-photographed' realm, or erase the distinction between them. Once the shutter has closed the moment remains forever trapped in a permeable membrane to be negotiated only by the present.

When I look at an object it is not just that I become conscious of it — it becomes my consciousness

The thing and the recording of the thing are distinguishable but not distinct

As with all permeable membranes, there is a flow across the walls. In our daily lives we are surrounded by numerous devices that, in one way or another, are designed to record aspects of reality: answer-phones, video cameras, tape recorders, CDs, DVDs and VCRs, etc. Each technology, whether analogue or digital, modifies some matter (magnetic particles, silver halides, vinyl, electronic impulses) in accordance with the reception of certain frequencies of energy that we can later retrieve. Beyond such mechanical descriptions, however, lies a galaxy of questions of the kind that continually pre-occupy media-theorists. To what extent does the recording constitute the thing recorded? Is the recording simply a mechanical facsimile that remains independent of what was recorded? Being a trace of something that happened, does it become an autonomous 'text'? Or is the reverse true; that the 'text' is a relic of the author, in the theological sense of the intimate fragment, through which something of the author lives? Does the 'text' in fact bring the author back to life (or negate the concept of death)? Might the old romantic notion of immortality gained through creativity have some validity? These questions will remain unresolved so long as an analytical methodology masks the thick membrane of the imagination in which objects and thoughts, matter and energy, flow reciprocally and without finite bounds.

Energism: Capturing the soul

We have already proposed in previous sections what might be called an 'energistic' view of reality in which all phenomena are considered in terms of the formation and transformation of energy. Proceeding from the idea that all things, including living things, are complexes of organised energy, we have claimed that the recording of an event traps some energetic aspect that is stored and recovered from a material substrate. And, since all things must include their consequences, the recording must be considered a part of the thing it records. When we are amazed by stories of 'primitives' who recoil from having their photograph taken lest their soul be held captive, we may now be more sympathetic and recognise the sense in which those less conditioned by materialism see the threat. Such minds may be more alert to the ecology of energy surrounding a living person that constitutes what some might call the soul. In this perception, any mechanical device that visibly traps an energetic dimension of a person would be potentially menacing since it can be removed from their jurisdiction. Taking such responses seriously, we are forced to reconsider how complex the interference may be between the mechanical and the psychic.

One day, forever

The ephemeral affects of organised energy can apparently withstand reduction to mechanical form without complete loss of emotional integrity. Think of the footage Abe Zapruder shot of the assassination of Jack Kennedy which ghoulishly evokes the events of a few seconds in Dallas in 1963 through a few hundred grainy 8mm frames. The energetic imprint left on our minds is still vivid: the bright, sunny day, the shaky camera work, the waving of arms and flags, the silent trajectory of the bullets and the jerking head as they hit their target. All these we feel, not as witnesses to the original scene, but through what the mechanical apparatus induces in our imagination. Thus something of the energy of an event transcends the mechanics of its recording to ripple outwards, indefinitely, through successive screenings.

Privileging quantity over quality

The conveyance of human drama through recording apparatus exposes the inadequacy of the mechanistic explanation of life and consciousness, and resuscitates something of the vitalist approach. Although the material content of the Zapruder film can be precisely quantified as matter (in materialist-mechanical terms), this gives us no useful information about its qualitative effect when watched. In the same way, we may be able to quantify precisely the mechanics of life or consciousness (DNA, RNA, neurons, etc.) without being able to extract anything meaningful about the quality of livings things that result. This is because, in abstracting complex events into computable terms, we necessarily have to omit qualities in favour of quantities — qualities are not computable. To compress the analogy, the mechanists believe that examining the material of the film can account for the emotions it arouses.

The eternal performance

Something, it seems, lives in the imaginary membrane: something that cannot be reduced to the medium, but nevertheless is contained in and recovered from it. When the twenty-one year-old Elvis Presley went into a Nashville recording studio on January 10th 1956 to record *Heartbreak Hotel* the session lasted about three hours. The song itself, considered by some one of the most erotic and haunting pieces of sound ever recorded, lasts about three minutes. It became his first US number one and has subsequently been heard countless millions of times. The work of a few hours, captured on magnetic tape and wax grooves, swells out across time, cultures and generations to be repeatedly recovered

A displaced statue of Lenin.

Trapped energy

What made Elvis Presley attractive to millions of fans; what makes an adolescent male figure attractive to Michelangelo Buonarotti; what did the Ipana marketing department want us to want from seeing the woman, her child and puppies? It is undoubtedly something of their specific energy, their 'charge' (as Gombrich puts it) which becomes embodied in the record, statue or photograph. And, being so embodied, it is trapped for as long as the medium lasts. We might not know the precise context in which the Palaeolithic rock paintings were made, or their meaning to people who made them, but this does not prevent us recovering their energetic vigour in our time.

Imagination, technology and desire in the image

Imagination, technology and desire meet in the typical photographic image of the toothpaste advertisement. The apparatus required to pose the shot, capture the light and mass produce the significant arrangement of matter (half-tone ink on paper, or gelatin on glass) induces the appearance of a magical state whereby something is both present and absent (the figures, their perceived state of being). Its purpose is to satiate a need and create a desire (later sublimated to promote the toothpaste, the preservation of genius or the instantiation of beauty). "Ipana dental care promotes healthier gums, brighter teeth." runs the copy that, in itself, encapsulates technology, imagination and desire — the science of dental care recovering the lost energy of youthful vigour. The vivid photographic image of brighter teeth and the desire for healthier gums leads unambiguously to an enhanced state of existence for the user.

Imagination and technology

The apparatuses for recording, capturing and displaying reality according to consensual norms are increasing in resolution, becoming cheaper to buy and easier to use. Yet their basic operation remains constant; they convert the transient energy we perceive into a semi-permanent trace that we can store and later retrieve through the effect induced in our imagination. It is clear that we cannot isolate active technology from imagination by devolving it to nuts, bolts, laser beams, wires and magnets, while restricting imagination to somewhere in the nervous functions of our brains. If we ever believed in such isolation it is because it was an presumption that was incompletely thought through. The mechanical, material techniques that we develop to trap reality form a substrate that only finds full expression in negotiation with the human nervous system. Technology and imagination in this sense are continuous, to the extent that they may appear to be the same thing seen from two different points of view. Such a relativist approach liberates new possibilities for synthetic thought.

Freedom from oneself?

Section Nine

Pornograph

But her long skirts hindered her, even though she held them up at the back, and Rodolphe, walking behind her, glimpsed — just between that black hem and the black boot — the delicacy of her white stocking, like a snippet of her nakedness.

Gustave Flaubert, *Madame Bovary*

Erotic gravity

Early relics of human culture often appear to represent sexuality as something integral to other strands of social existence such as family, religion and fertility. The carving in the Dorset hillside known as the *Giant of Cerne Abbas* is an example of an ancient and highly public depiction of human erotic energy. Today such an explicit public representation of male arousal would be unacceptable. In general, the depiction of sexuality is now the monopoly of the state and devolved to legitimate and illicit pornographers subject to strict legal restrictions. This diverts the representation of human sexuality into an occluded, self-contained tributary running parallel to the main flow of social activity. Despite this masking the gravitational pull of the libido exerts an irresistible force. Michel Foucault has argued that sexuality has never been erased but re-encoded as part of a different social order as deep and dangerous currents of libidinous energy create turbulence under the apparent calm of the surface. There are few aspects of our lives that do not fall under its influence.

Pornography and technological change

The developmental course of apparatus for capturing, storing and distributing audio-visual information has been influenced at several critical junctures by the gravity of (usually male) erotic desire. Without the intervention of pornographers, for example, it seems likely that we would be using the more compact and reliable Betamax rather than their preferred format of VHS. The initial appeal of photography, stereoscopy, cinema, domestic video, satellite & cable, CD-Rom and the Internet as male enthusiasts' hobbies can be linked to a covert erotic dimension as new ways of distributing the eroticised body are promoted. In each case, as a peripheral technical hobby becomes institutionalised, its earlier appeals are submerged but remain as an erotic undercurrent. Base as the erotic impulse may seem, it conjugates certain technologies with the body and in so doing produces a kind of artificial life.

The difference between pornography and erotica

Sexual representations tend to be classified as either 'pornographic' or 'erotic' — a distinction usually made by the degree of apparent explicitness. Yet trying to make this distinction on the basis of content is ultimately confusing. One can point to many examples of images in pornographic publications that are less explicit than some perfume adverts from mainstream magazines. As we see with the example

144

from *Madame Bovary* cited above, the eroticism of the body is not governed by the degree of exposure but the expectations (and tolerance) of the market; the contents of particular sexual representations are influenced by the needs of those who buy them. Pornography (etymologically: the writing of prostitutes) is a commodity that the pornographer sells and, as such, any product of this particular market can be considered pornographic — regardless of its content (the market can be defined quite precisely as consisting of certain publishing concerns and distribution networks). Pornography certainly connotes a particular lack of subtlety in the commodification of sexual representation and the content of pornography, its distinctive 'rawness' and super-artificiality, may be due to it being more visibly (and brutally) exploitative because of the exploitative nature of its production infrastructure. The immense profitability of the trade comes from its low investment/high return strategy. Erotic material, on the other hand, may be sexually explicit, but it is not pornographic in so far as it is produced and distributed outside the pornography business. Erotic material may require more intellectual investment and may be richer and subtler in content because it needs to satisfy the demands of a different or wider market — a market more concerned with quality of expression than monotonous arousal. Erotica is still produced commercially but not necessarily as part of the pornography trade and not with immediate arousal as its sole imperative. In the long association between sexuality and art, many artists and writers have produced erotic works that are regarded as having artistic merit to meet the proclivities of particular clients — Anaïs Nin is supposed to have produced many of her best known works at a dollar per page.

The manufacture of pornography

Given the extended energy of human sexual desire, it is not surprising that lens technologies are irresistibly lured towards the artificial supermen and women who contribute to the super-artificiality of pornographic products. Lens technologies produce rapid and reliable metonyms of human desire, and are inevitably attractive to the manufacturers of pornography. The metonymic conventions that make photographs intelligible are deployed to amplify, extend and idealise the aroused body. Film editing suggests unmanageable feats of endurance and patience where the connective material of life is eliminated in a super-scene of desire in which there is no respite from one single obsession. Just like the 'real' movies' it's little wonder then that, with so little of the quotidian aspects of life on screen, porn-films can induce the boredom of un-modulated energy. In pornography, the camera enables limited transfer of a specifically sexual energy (conspicuous displays of animation are required of the actors to offset the limitations) that can be retrieved by the user at will, without the usual chores of dating, mating and hating.

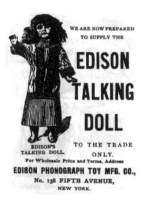

The mechanisation of pornography

The erotic possibilities of cinema technology were present at the outset, and arguably in Thomas Edison's mind when he spoke of moving picture machines prior to their invention. His fascination with artificial life can be seen in the well known letter of 1887 which speaks of his ambition to manufacture sounds and images of opera stars long since dead. Equally well-known is his depiction in, and obsession with, Villiers de l'Isle Adam's *L'Eve future*. As Edison was realising his erotic ambitions with Mina Miller in frivolous games he was exploring new ways to make pictures move. The ambivalent erotica of Muybridge's studies of naked men and women, which Edison collected, provided the perfect opportunity to realise this desire and has been linked with his interest in making these pictures move in the Kinetoscope. In any case, the notion that 'inert' matter could be endowed with erotic charge was certainly plausible (see note 8).

What does pornography offer?

How is it that 'lifeless' matter such as an arrangement of silver halide crystals on celluloid, or colour values of screen pixels, or ink dots on paper can apparently arouse some of the strongest aspects of human desire when in negotiation with the nervous system? What lies in the means of depiction, in the sediment of the inert mechanical matter, that has the power to fuel one of the most profitable markets in the world? Surely it is the spectacle of specific human energy as manufactured data and the purchase of energetic continuity with an object of sexual desire. Pornography is a precise example of the operation of representation as well as the bodily effects of imagination overriding perception. In the exchange between pornographic object and viewer, the arranged matter is mis-recognised for the desirable other — with consequences for the conscious flesh (see note 9).

The poverty of pornography

If sexual pleasure is, at least in part, linked to the overall increase in energy flow usually produced by conjugation with the energy of another (or by the thought of it), and if the condition that arises through such contact satisfies, at least in part, the desire for unification with another, then pornography ultimately fails to satisfy on both counts. The sexual energy retrieved from the pornographic image, in the first instance, is normally simulated by actors and its effect on the viewer, though instantaneous, is consequently short-lived. In the second instance, the continuity between the viewer and the subject of the image is limited by the fact that the subject is accessible only through the imagination and the representational device. The sensory data is limited to an audio-visual stream which is normally monotonous, whilst the other senses of taste, smell, touch and the desire for response remain neglected. So while the viewer and the subject of the image are actually continuous (in that all things include their consequences and consciousness extends to its object) the limited energy data available forbids full unification and, hence, leaves room for more desire.

Warming sign

The subversive pleasure of enjoying ourselves

In recognition of the limited appeal of raw human bodies as a source of energetic excitement the human actors in erotic images are often supplemented by the staging of fantasy scenarios that use provocative clothing, music and lighting. Moreover, such embellishments are often the very signifiers of the pornographic process, energised by their assumed subversiveness and thereby avoiding quotidian depictions of anatomy. Pornography usually presses on the boundaries of a given society at a given time and, although human sexuality cannot be repealed, material evidence of its presence is widely prohibited. But by leaving the unadorned and apparently un-sexual body to the Life Room, the pornographers engage two of the most irresistible agents of human behaviour — the erotic and the taboo. From the very start the cinema found its revenue in the erotically charged atmosphere of a darkened dive where images of boxers stripped to the waist were accessible (a taboo for American women who were not allowed into the matches). Why else did the stag film drive the market into an attractive disrepute at the very moment when its first audiences became bored with phantom rides and bad boy pranks? Pornographic images, then, are not simply records of sexually desirable people, but depositories of amplified libido, further amplified by their notoriety.

Pornography: Imagination, technology and desire

However they appear, whether raw or highly mediated, the effect of pornographic images is secured by a remarkable formal consistency driven by one aim — the relay of specific human energy through representation. Pornography provides a very clear instance of where imagination, technology and desire become mutually implicated. Connecting the fields of desire and technology — the wants and the solutions — is the expanding presence of human imagination as something that exists beyond the human brain. In our imagination — the arena in which division and continuity co-exist — the fragmentation of language can be temporarily reconciled, just as the division between ourselves and the world is reconciled in (imagined) sexual union.

Transcendent imagination

If the expanded imagination is understood as a membrane in which conflicting desires arising from the world are reconciled and technologies of human extension are conceived, then it follows that what is 'inside' cannot be privileged over the 'outside'. The mind in this model acts as a synchronous transmitter and receiver that is functionally conditional on giving expression (gestures, speech, communication) and receiving stimulation (movement, sound, language). It is in this sense that we may talk of a transcendent imagination: one that is not reduced to (internal) neurological impulses in organic tissue but includes the sum of (external) reactions and effects associated with the imaginative experience. Given that no boundary can be drawn around it, we have to consider the imagination as a thick film that has been arbitrarily spliced for the purpose of *post-hoc* analysis.

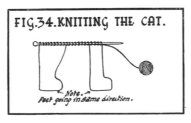

Diagram

Encoding the imagination

The compulsion to 'externalise' or 'express' thoughts tends to result in the development of complex means by which they can be made permanent or portable. The techniques that have evolved for achieving this obviously include language (speech) which, it has been argued, is the primary vehicle for encoding and communicating thought. However, language is not only seen as a way of encoding mental activity but also as being synonymous with it since a great deal thought occurs in the form of language. As with all encoding, the notion of encoding thought in language implies a translation — turning one thing into another without significant loss of content. This rather mathematical, linguistic approach, however, allows little room for the possibility of extra-linguistic thoughts that might precede language's attempts to encode them. For example, we are daily aware of visual, abstract or emotional ideas that resist expression as speech. The inadequacy of language becomes more acute as we consider that imagination contains those aspects of our mental life brimming over with flashes of insight, intuitions, feelings and visions. Of course, such ill-defined concepts seem particularly messy if you are trying to represent human consciousness as a logical symbolic system (as we suggested earlier some AI advocates attempt to do) since it implies a strata of human consciousness impervious to symbolic representation.

Other than words

Evidence of the thickness of the imagination can be seen in the fact that, despite the (word-based) human language facility being highly evolved and flexible, it has not displaced other modes of representation or communication. For example, we still invest human energy in painting, drawing, singing, photography, spatial construction, filmmaking, sculpting, and so on. While these forms of expression may be considered to have linguistic characteristics in that they appear to have rules of context, grammar and syntax, they contain information that defies linguistic representation. For example, it may be possible to construct a homologous 'language of music' by the imposition of rules and structures but it is not possible to translate music into text. The musical experience is, therefore, extra-linguistic and exemplifies a mode of imaginative expression that resists symbolic codification.

Translating the imagination

Although we understand that the translation from experience into cultural artefact is not a complete equivalence, nevertheless this does not inhibit us from creating new works that attempt to materialise aspects of mental activity. Nor does it limit the demand for objects that appear to do this reliably. The idealised concept of translation implies a symmetry that denotes transference from one place to another, and it is in this sense that imagination is (in a limited way) moved through the membrane from the 'inside' of the person to the 'outside' in the world when it is materialised. As such it does not cease to be imagination when materialised, but merely imagination that has been translated, or moved to a different place (albeit somewhat changed — even reduced — in the process).

Recording the imagination

Tourists who experience London's *Big Ben* through a viewfinder must expect the resulting images to invoke some of what they experienced from seeing the real thing. However, we are all aware of the deficiencies of any recording system and on looking at their work we expect to receive only a partial sense of what the artist or photographer experienced. No one imagines, for example, that Van Gogh's late landscapes offer a complete description of his environment. Just as the encounter with the sexual other through pornography is ultimately incomplete, so poems about, or paintings of, the trenches in the Great War may be moving but not dangerous.

Meaning is negotiation

The reception or reading of a translation of human imagination (linguistic or extra-linguistic) is extraordinarily complex and subject to the vagaries of cultural context. It could be argued that the articulation of this complexity is itself well beyond the scope of language and even beyond our understanding. Indeed, the very pleasures of a work of art may depend on the degree to which the limitations of language become evident, where moments of unity are rescued from the everyday fragmentation of language. Nevertheless, language, especially in its written form, is a relatively reliable means of transmitting human imagination, although as with film, the implicit content is always far greater than explicit content. Meaning is not explicitly carried in the material but implicit in the negotiation between the material and the imagining human nervous system. This would account for the relative reliability of linguistic communication, as well as its widely recognised variability.

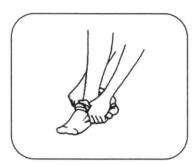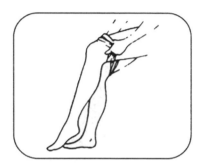

Non-linguistic negotiation

Decoding the imagination across time

In many kinds of art the measure of the value of the work may lie in the degree to which it reliably invokes rare sensations, desires that language can only try to name. Given the human sensitivity to (libidinal) energy it is not surprising that the focus of much art is on the dynamics of erotic attraction. In *Madame Bovary*, Gustave Flaubert's description of the besotted Léon observing Emma, provides a compelling moment of self-recognition in the reader:

> First they had a few rounds of trente-et-un; then Monsieur Homais would play écarté with Emma; Léon, standing behind her, gave his advice. Standing with his hands on the back of her chair, he gazed at the teeth of the comb thrust into the coils of her hair. Every time she reached out to play a card, it lifted her dress on the right side. From the coiled mass of her hair, shades of brown flowed down her back, until, fading away gradually, little by little, they ended in shadow. Her dress, as she sat back again, spilled

over both sides of her chair, in ample swelling folds, that reached right down to the floor. When Léon sometimes felt it under the sole of his boot, he stepped backwards, as if he had trodden on something living.

Gustave Flaubert, *Madame Bovary*

This description of a man in love, despite its coyness, vividly transports a moment from nineteenth century rural France into the mind of the contemporary reader. This transmission of desire through time and space, and the continuity between our mind and Flaubert's, is echoed in Vladimir Nabokov's *Lolita* in which the presence of Flaubert is clearly audible:

I let my hand rest on her warm auburn head and took up her bag. She was all rose and honey, dressed in her brightest gingham, with a pattern of little red apples, and her arms and legs were of a deep golden brown, with scratches like tiny dotted lines of coagulated rubies, and the rubbed cuffs of her white socks were turned down at the remembered level, and because of her childish gait, or because I had memorised her always as wearing heelless shoes, her saddle oxfords looked somehow too large and too high-heeled for her.

Vladimir Nabokov, *Lolita*

In this evocation of doomed desire, the narrator Humbert objectifies the subject of his fantasy, Dolores, into a composite of flesh and commodity. Like Flaubert, Nabokov constructs a 'join-the-dots' constellation of recognisable detail that requires the energetic negotiation of the reader to draw in the lines for completion. Almost like a film camera on a crane, Flaubert and Nabokov both 'shoot' these scenes by scanning the women (through the lens of another man) from head to toe: lingering on details (the lifting dress, the scratched limbs, etc.) that might transfer a specific energetic pattern to the mind of the viewer by a process of mis-recognition — we mistake the marks we perceive on the page for objects in our imagination. The process of encoding imagination, in these cases, assumes a corollary process of decoding that, if accurate, conveys the sensation largely intact. In such acts of extended consciousness, the imaginations of the writer and reader are conjoined momentarily across a void of time, space and cultural experience.

153

Conjoined momentarily across a void of time

Technology as extended imagination

Is our own capacity for imaginative thought impoverished by improvements in the fidelity of encoded experience? The possible means by which human thoughts have been recorded, stored and conveyed have rapidly expanded in recent history. At the beginning of the last millennium the ways of encoding imagination were limited almost entirely to speech, writing (including illuminated manuscripts), music, painting, textiles, metalwork and masonry. At the beginning of this millennium the number of available recording devices and techniques for the purpose almost defies quantification. Would it be fair to conclude from this quantitative increase in information that we are now more effective at encoding human imagination and experience than our predecessors were? If imagination can exist outside of our physical bodies and be consumed like any other commodity, will its relative value be diminished as it becomes more abundant? The average visitor to an eighteenth century Salon would expect to spend an hour or more in front of a painting, discussing, unravelling meanings, linking narrative and symbolic strands arising through the contemplation required by religious or historical art. Is the same concentration and devotion shown to contemporary art-works? Do we now treat acts of imagination as we treat drinking water? Could it be that we are now, more than ever, storing our imaginative resources remotely so that they can be retrieved more conveniently — 'on tap'?

Technology distributes imagination

Cultural artefacts from earlier times, such as the *Giant of Cerne Abbas*, often appear crude to our contemporary sensibilities as we look at them in isolation from their ritual function, stripped of the full richness of their utility and local meaning. With their relative simplicity of form, early human objects may have left more to the imagination than we are generally able to appreciate today. There is little evidence that human desire has changed over thousands of years, but increases in technological sophistication may have been won at the expense of our capacity for imaginative enchantment with the Universe.

The Giant of Cerne Abbas

Unifying technology

In this section we have argued that the discourse of energy transfer is highly visible in the production and consumption of erotically charged representations. Connecting these fields of desire and technology — the wants and the solutions — is the expanding presence of human imagination, procuring for itself moments of unity from the fragmentation of quotidian consciousness.

Unifying technology

Section Ten

PROPOSITION:
The categories of logic, reason and the binary are absorbed and transcended in the postdigital membrane

Zeno's paradox

Zeno's paradox

One problem has always dogged those philosophers who advocate a rational Universe made of divisible units. Many of Zeno of Elea's paradoxes demonstrate how the transition between one state and another is composed of infinite steps so that, by the force of reason, no two states can be divided. How, for example, when a person is thirsty do they transmute the desire to drink into the action of lifting the glass? If an arrow is static at each distinct moment when it flies, how does it ever reach its target?

157

Zeno's return

Although the well-known paradoxes of Zeno are for most people nowadays considered as little more than logical oddities, his reason for advancing them four hundred years before Christ had a more profound purpose. He sought to show by example that using reason to distinguish between things was absurd. In the world of experience, one could never absolutely define a point where one thing ends and another begins. Zeno's paradox reminds us again that in our unmediated consciousness there is no distinction between things, and differentiation is no more than a consequence of language and the operation of our senses. The postdigital migration of the body, and its fusion with the technological, resuscitates Zeno's position. Reason dissolves as the sensory world blends with the 'real' one. This causes us to reject the reduction of continuous reality into binary (or any other) discontinuous segments.

The Actor's paradox

The Actor's paradox

Actors are able to convince the audience that they have levels of competence (or incompetence) that they do not really possess.

Imitation and Artificial Imagination

As we argued in Section 5, the failure of computational AI — a scientific project that arose from a human conceit — is but a symptom of a more profound malaise of the seventeenth century project of reason. During the 1670s a Dutch artist, Cornelius Gijsbrechts, working in the Danish court of Frederick III was developing an astonishing technique for painting trompe-l'oeil still-lives. Gijsbrechts specialised in the genre of illusionistic painting in which the various debris of life — letters, quills, prints, combs — are rendered with such obsessive detail that they appear, at first glance, to be actually pinned to a wooden board. Despite the intervening 300 years, and the invention of colour photography, the deceptive effect of these works is surprising when seen at first hand. What is more, the visual puns and inter-referential meanings in Gijsbrechts' paintings anticipate the best works of the mid-to-late Cubist period with their paintings of paintings, simulations of surfaces and arrangements of everyday objects. One work in particular of 1670-72 depicts the reverse of a framed painting showing the nested wooden frames and the primed back of the canvas. This painting, which itself is unframed, is displayed leaning against the wall rather than being mounted vertically. The immediate effect on the viewer is unsettling and mildly euphoric as one takes it to be what it purports to be — a reversed painting stacked against the wall. Then comes the almost instant realisation of the illusion and the subsequent moves to confirm the actual state of affairs. What Gijsbrechts offers us is an image that conforms precisely to the description of art (indeed, of all representation) proposed in this book — the skilled arrangement of matter and energy (paint, canvas, light) so as to add significance by invoking the presence of something that is also absent. This painting demonstrates that what we think we perceive we actually imagine. In this case of convincing imitation, where one thing pretends to be another (the organic world is full of examples) we are reminded of artificial intelligence and artificial life and the skill with which coders and researchers may be able to arrange matter and energy (code, switches, electrons) so as to suggest the presence of a kind of absent human intelligence. As a form of entertainment, imitation may be highly amusing, it may even have military applications (the camouflage introduced during the Great War is thought to have been inspired by Cubism), but it is ultimately no more than a ruse to fool our perceptions.

What is more interesting is the failure to recognise the persistence of real human intelligence in all matter we deliberately transform; a failure caused by the limited conception of intelligence as being something confined to the brain.

Gödel's paradox

Gödel's paradox

Kurt Gödel was able to show that in arithmetic not all propositions could be proved even if they were true. This is taken to mean that some truth lies outside the system under interrogation, which is a paradox of logic that is for some only resolvable by endowing the observer with a separate status.

The vogue for binarism

The vogue for binarism (that is the reduction of complex social realities to oppositional structures such as raw/cooked, male/female, nature/culture) is widely regarded to have been initiated by the work of the anthropologist Claude Lévi-Strauss and taken up by the Structuralist school that emerged in his wake. As a means of classifying and analysing social reality, structuralism itself claimed ascendancy from both the linguistics of Ferdinand de Saussure and the pioneering work of the American communication theorists of the 1940s such as Claude

Shannon. Given the optimism at the time about the emerging fields of mathematically inspired computer science and comparative cybernetics, it must have seemed obvious to link computer modelling and social organisation in that they apparently shared a common binary structure. The consequent tendency to try and reduce reality to logical formulae, and hence binary digital code, finds expression in the todays fascination with the mirage of digital life and consciousness. It is as though the digital age is the apotheosis of humankind, the conclusive mathematisation of reality. We are presented with the dubious assumption of the inevitable 'digitisation of society' that will incorporate entertainment, commerce, intelligence, life, consciousness, sex, communication and medicine.

The Postdigital

The widespread enthusiasm for the computer is currently running high and ever-greater things are expected. Undoubtedly there have been changes in the way we talk, think, fight and trade that are linked to the demands of computer technology. But as the conceptual paradoxes and the technical limitations of digital computers become apparent there will also be disappointment and frustration. The Universe is not digital but it can be digitised. Consequently, digital machines will never be independently conscious (in the way we currently understand those terms in all their full constitution), and living things will never be reduced to the mechanical (as we currently understand that term). For the anthropological notion of the 'binary' to be compatible with the computational notion of 'binary' the two terms must be exclusive. '1' or '0' can only be '1' or '0' — there can be no ambiguity or irrationality. However, the profound opposition between male and female, for example, is never clear cut since there are infinite degrees of masculinity and femininity. This does not negate the utility of either concept in as far as they are applicable to our daily experience, just as one does not deny the utility of 'night' and 'day' because there is no absolute boundary between them. As Zeno's paradox insists, the areas of transition between states are parts of each state they become. It is not possible to represent infinite gradations on a digital system because, even with an unlimited amount of memory and computing power, one always has to impose binary divisions, which may be reasonable, but will always be arbitrary

State transition

The postdigital computer

One attempt to accommodate the complexity of experience in a digital machine is the 'postdigital computer', although such machines are currently only a theoretical branch of nanotechnology research. In *Design of Basic Elements of Digital and Postdigital Computers Based on Quantum Mechanical Investigation of Fullerene and Photoactive Molecules* published on the Foresight Institute web-site there is documentation of an attempt to model the activity of neurones by offering a range of logical states greater than either 'yes' or no'. The researchers describe their proposals for the types of logical structures that could be offered in nanotechnology computers:

> Complete set of sixteen MIs of two variable logic functions (for example: OR, AND, Implication, Equivalence, Difference, etc.) was designed and also proposed using MIs of two variable molecular logic function initial basic sets: {OR, AND, Negation} or {NOR} and, or {NAND}. We have described in more detail the designed MIs of: a) two variable logic functions OR, NOR, AND, NAND (two sets: one designed from planar molecules and another — from fullerene molecules). Converse Unitary Negation-1, Converse Unitary Negation-0, Unitary Negation-1, Unitary Negation-0, "0" and "1" Matrix= Constants; b) three variable logic functions AND, NAND, OR, NOR analogs; c) four variable logic functions OR, NOR, AND, NAND analogs, d) molecular cell that simulates one of Life figures, e) summator of neuromolecular network that simulates sigmoidal behaviour of artificial neurone.

However, even if it were possible to build machines that operated according to the extended logical functions described above, they would still have to conform to the constraints of logic which, in the context of the postdigital, sets limitations on the field of thought. A truly postdigital computer will work in ways that are both logical and non-logical (just as we do).

Transcending binarism

The postdigital era has been foreshadowed, not only in stubborn sense of Zeno, but also in Roland Barthes prescient remarks about structuralism written in the early 1960s:

> In fact, and to conclude briefly on the question of binarism, we may wonder whether this is not a classification which is both necessary and transitory: in which case binarism would also be a metalanguage, a particular taxonomy meant to be swept away by history, after having been true to it for a moment.

Roland Barthes, *Elements of Semiology*

This is a timely reminder that the digital age is transitional and quite possibly short-lived. Whilst the power of digital processing to model and control complex systems seems beyond challenge, we will overlook its weaknesses. These lie in the inherent inflexibility of the binary encoding of information which is limited to 'all or nothing', 'on' or 'off'. The terms of a binary expression must be mutually exclusive. Any information that is digitally encoded is either inevitably degraded or must succumb to logical encryption. Higher sampling and bit rates may increase fidelity beyond the limits of human perception, but can never capture all the analogue information. Any idea, calculation or problem one might want to input into a digital computer has to be encoded into a logical form — it cannot be stated ambiguously or irrationally. While one can instruct a computer to perform some extremely complex tasks, even program in levels of unpredictability and randomness, one will not overcome the fact that the information has been abstracted into a logical form. All digital computers use mathematical logic to

163

encode and process information. Since there are many aspects of reality that resist rational encoding, the digital computer will never transcend binarism. Consequently, it will only ever simulate complex human experience to the satisfaction of language.

The Postdigital Membrane

In the postdigital membrane identifiable parts are not reduced to oppositions and analysed as dialectical constellations but considered as continuous with each other. However, in order that we can enter into useful discussion we can retain the distinctions only as long as we also acknowledge their continuity. To achieve this apparently impossible mental task we adopt the metaphor of the membrane. For example, as we have argued, the notion of human imagination can extend beyond the immediate locus of the human body into all of reality by permeating the imaginary membrane that separates them but also specifies them. In general, we can conceive of 'things' (distinct perceptions) as being both connected by and separated by permeable membranes that exist both in our minds and in reality (since we cannot divorce one from the other). While a membrane is inevitably a constituent of each thing it lies between, it also has its own constitution in as much as it regulates the flow of energy across its surfaces. This model does not preclude divisions such as nature/nurture, order/disorder, reality/imagination, etc. but it does insist that they are distinctions that we impose upon the world. As such, they do not have *a priori* status in themselves — they don't appear until we produce them. Having produced them it is legitimate for us then to question and reconfigure them, especially if they no longer seem useful or relevant.

Imagination, Technology and Desire

We have reached for an understanding of these three terms: imagination, technology and desire and, perhaps, an appreciation of how futile it can be to separate them. Each term can be likened to a neighbouring political state on a large mass of land; they may have their own boundaries, languages and regional identities but these are essentially abstract constructions of human mentality imposed on the terrain. The earth below and the air above remain unbroken.

The Chinese symbol for
Yin and Yang is not a
binary symbol
because each
side carries
something
of the other
— the
opposites
are not
mutually
exclusive,
they are
essentially
analogue, and
complementary

The Chinese symbol for Yin and Yang is not a binary symbol because each side carries something of the other — the opposites are not mutually exclusive, they are essentially analogue, and complementary

Notes

1. Back to the future

It may well be that we project forward the satisfaction of our desires into the future and by doing so often ensure that those desires are met. Will we ever become invisible, travel through time or read people's thoughts? Perhaps we only dream those fantasies because they are technically feasible within the bounds of natural law? To what extent did Leonardo's crazy little drawing inspire helicopters? Or to reverse that question — to what extent did the present existence of helicopters influence Leonardos drawing, or our interpretation of it? We dreamed of submarines, moon-walks, horseless carriages and artificial limbs — now we have them. Do we only want what is possible or does the possible determine what we want?

2. The human membrane

Something of the human as membrane metaphor was suggested to N. Katherine Hayles in *How We Became Posthuman* by Norbert Wiener in his book, *Cybernetics*. Talking of the flow of sexually attractive chemicals between bodies. Hayles writes:

> ...the pheromones that guide insect reproduction are general and omnidirectional, acting in this respect like hormones secreted in the body. The analogy suggests that external hormones organize internal hormones, so that a human organism becomes, in effect, a sort of permeable membrane through which hormonal information flows.

Hayles argues that Wiener did not embrace the implicit consequence of this analogy, that is the dissolution of the subject-object boundary. It may be that we are now ready ourselves to accept it. *The Post-Human Condition* demonstrated that it is no longer feasible to think of the human being as a separate 'thing' within the world. It argued that we are not isolated 'things' that exist within an environment, or in conflict with it, but part of and continuous with it. In the case of a human, the permeable membrane (literally and metaphorically) separates us from and makes us continuous with the world.

3. Diminished perception

The energetic transmutation of intellectual matter from a diffuse to a cohesive state is achieved at a price. As images drip-feed data into the system with the same effortless ease that we might absorb our mothers milk, there is a process of artifice and human reduction that diminishes the breadth of our perception, even in our most intimate relationships. Yet love, like language and history, is re-configurable. While we have the opportunity to engage with rich multidimensional platforms of action, we are forced to address the limits and constraints that we impose on our own life in favour of an imaginary future.

4. To ask why or how

To ask why or how something happens (as we do in this book) is to seek to de-fragment the world. To connect a cause and an effect is to close a fracture, negate a blind-spot, heal a wound in the fabric of reality. Those who want to create intelligence in a machine are driven by the hope that the production of intelligence in an explicable medium will mean that intelligence itself succumbs to explanation, explanation being the the connecting of apparently diverse and unconnected aspects of reality.

5. The Theme Park of progress

This is not to claim that either the painting, the blackboard or the candle were the sole determinants of a new self consciousness. To do so would be to deny the other changes in instrumentation, transport, mathematics and language itself — not least in the realm of philosophical thought. The acceptance and exploitation of a mediation of perception (for example, photography) does not necessarily occur at the same time as its technological manifestation. During the nineteenth century industrial processes and information technology altered the human perceptual and expressive apparatus. The storage and transmission of data by means of telegraph, the reliable duplication of objects and images by mechanical means, and the embodied intelligence in mobilised human beings travelling at (relatively) high speed in trains were all developments as radical and creative as the eighteenth century candle. Some, like the ticker tape telegraph that converted remotely sequenced digital codes into punched strips of paper, have been consigned to the theme park. The production of identical images using photosensitive materials are perhaps heading that way, whilst others, the railway for example, are still functioning as the best system we have. But to equate the persistence of a technology with a persistence of its effect is to propose a diachronic account of synchronic events. It is to isolate one technology as paramount, retrospectively casting it as the fulcrum it may not have been. Significant developments in the technology of art suffer from the same retrospective classification that demands a chronology of perpetual progress (improvements in technique, materials, perception, insight, etc.). This renders earlier works, and even the output of whole cultures, as deficient in one respect or another, or as steps on the way to something better. The consequence is an impoverished understanding of those aspects of past works that are difficult to quantify but which, nevertheless, provided their impetus at the time they were made. It also blinds us to the possibility that, in losing its radical social function as a means of social transformation, art has got worse rather than better. The Museum, like the theme park, is a repository for that which can no longer be used in its intended context

6. The assumptions of some Scientists

The renowned physicist John Barrow is a paradigmatic example of a 'hard scientist' in the way in which he expresses support for scientific method as a series of testable assumptions. He suggests that:

- There is an external world separable from our perception
- The world is rational: 'A' is not equal to 'not A'.
- The world can be analysed locally — that is, one can examine a process without having to take into account all the events occurring elsewhere.
- There are regularities in nature.
- The world can be described by mathematics.
- These presumptions are universal.

John Barrow cited by Lewis Wolpert, *The Unnatural Nature of Science*

Behind this mantra of scientific logic is a conspiracy of illogical reasoning. In the context of this postdigital account they are profoundly mistaken and cannot be left unchallenged:

- *There is an external world separable from our perception.*

As we argued in previous sections, although there is little doubt that we perceive the external world as being separate from us, we should not necessarily believe what we perceive. The world and our perception of it are inseparable.

- *The world is rational: 'A' is not equal to 'not A'.*

A tautological and recursive statement. Rationality is no more than a means by which we try to consciously impose order on the world and, as such, is vulnerable to the inconsistencies of language. Such a statement is true only in reference to its own terms, i.e. the terms of language.

- *The world can be analysed locally — that is, one can examine a process without having to take into account all the events occurring elsewhere.*

This is fine providing one does not forget that any process is an arbitrary idealisation of reality made in order to help us understand and control it. As René Thom declares in *Structural Stability and Morphogenesis* "In particular, the dimension of the space and the number of degrees of freedom of the local system are quite arbitrary — in fact, the universal model of the process is embedded in an infinite-dimensional space." Therefore, we must be careful in what we choose to ignore since Chaos Theory suggests that small forces can have large-scale effects in turbulent systems. Transferred to the domain of ideas (a turbulent system if ever there was one) this might suggest that one cannot extrapolate universal truths (large-scale) from local events (small-scale).

- *There are regularities in nature.*

The regularities in nature are consequences of our perceptual apparatus and consciousness (as is the case in other conscious creatures). Patterns and connections do not exist without being perceived as such. Scientists cannot objectively investigate an external reality since their perceptions and consciousness are part of the reality that generates those patterns and connections.

•*The world can be described by mathematics.*
Only the world that can be described by mathematics can be described by it — but that does not make mathematics into the world. Even Mathematics cannot describe its own illogicalities.

•*These presumptions are universal.*
In other words, they are beyond criticism — and we have shown they are not. The history of science is full of embarrassing examples of similar ahistorical claims.

7. A brief history of human energy

Much philosophy and myth-making in ancient writings concerns the optimisation and conservation of human vitality. The Asian practice of the Tantra, though varying from sect to sect, describes the focused channelling of sexual desire, pleasure and energy for the purpose of achieving ecstatic self-extinction. It works on the principle that correctly sharing or consuming the energy of others immensely enhances ones own potency, creativity and vitality. The idea that one can 'tap' the energy of another to restore one's own is common, even in our most revered texts. Ancient Hebrew legend tells in the Old Testament of how an aged king sought to revive his fading powers through non-coital consumption of a girl's 'heat':

> Now King David was old and stricken in years; and they covered him with clothes, but he gat no heat. Wherefore his servants said unto him, Let there be sought for my lord the king a young virgin: and let her stand before the king, and let her cherish him, and let her lie in thy bosom, that my lord the king may get heat. So they sought for a fair damsel throughout all the coasts of Israel, and found Abishag a Shunammite, and brought her to the king. And the damsel was very fair, and cherished the king, and ministered to him: but the king knew her not.

Kings, Ch.1, v.1, (King James)

In this story there are similarities with the folk-tale of the vampire, enshrined in

Gothic iconography since at least the eighteenth century. This creature can only maintain its viability through the consumption of (often virginal) human blood, an act that adds a vicarious erotic dimension to the acquisition of the vitality of others and which has been played out in numerous books and films. This motif is repeated in other myths, notably in the nocturnal activities of incubi and succubi fabled to sap the moral energy of women and men through illicit sexual preying. Blood has frequently been seen as a major source of human energy (because its loss leads quickly to death) and youth as a repository of vigour. Such is its symbolic value it has been frequently offered in the appeasement of fertility deities throughout time, ostensibly to influence the course of divine nature. Examples of this exist across world history: from early South American civilisations such as the Incas to Ancient Britain, Greece and Africa. These acts of bloodletting betray a consistent belief in the significance of human energy contained in youthful flesh whilst also publicly reinforcing the position of those who can authorise its conspicuous waste. In more civilised manifestations such dramas are reincarnated in sport and in some forms of art (dance) in which youthful energy is both displayed and dissipated. In recent times questions about the relationship between matter and vital forces came to be the examined in the domains of literature and science. Mary Shelley's protagonist, Baron Frankenstein, is reluctant to disclose the technique by which he restores life to dead matter, but the revelation of his discovery is described in the most energetic terms:

> I paused, examining and analysing all the minutiae of causation, as exemplified in the change from life to death, and death to life, until from the midst of this darkness a sudden light broke in upon me — a light so brilliant and wondrous, yet so simple, that while I became dizzy with the immensity of the prospect which it illustrated, I was surprised that among so many men of genius who had directed their enquiries towards the same science, that I alone should be reserved to discover so astonishing a secret.

Mary Shelley, *Frankenstein*

The technique of restoring the monster's life has latterly been represented in the cinema as an insurgence of electrical current, often directly sourced from lightning, as if to emphasise the divinity and excess of the energy required for life. In Britain, during the decades following the publication of *Frankenstein*, science was in the process of being organised into a socially regulated structure. This was in order to wrest it from the dilettantes and individual enthusiasts who had practised Natural Philosophy in the previous two centuries. It was not without resistance as Alison Winter points out in *Mesmerized, Powers of Mind in Victorian Britain*. She charts the cultural terrain that was invaded in the mid-nineteenth century by 'animal magnetism' as she describes how the period 1820 to 1840 spawned a proliferation of Scientific, Academic and Medical institutions and related legislation. Against this background of specialisation and regulation, those of a progressive or radical persuasion embraced the new concept 'animal magnetism' that was to become known as 'Mesmerism'…

…largely because it promised to address many of their questions about the relationship of physical forces to life. To what extent should scientific accounts emphasise matter, as opposed to vitality and spirit? Was the organising principle of life inherent, or was it breathed in by the Creator? Was there anything essentially "vital" about living things? What was the relationship of electricity and other physical forces to the mind?…Animal magnetism seemed to have the potential to address all of these questions, and to illuminate or to intensify connections between them.

It is clear that such questions remain with us today. The notion of animal magnetism was extended, albeit in differing forms, into the early twentieth century by those such as Reich, Bergson and Freud whose ideas owed a great deal to the practice of hypnosis, derived directly from mesmerism. Reich and Freud used the techniques of hypnotism regularly as part of the psychoanalytic process. Wilhelm Reich, a former student and colleague of Freud, gradually extended the idea of a living energy to a kind of sexual light he called orgone. In his 1942 book *The Discovery of the Orgone* he outlined its major characteristics and the methods by which it could be collected, stored and used. In the 1940s he built devices for this purpose called orgone accumulators which, he claimed, allowed the healing of disease and the negation of harmful radioactivity. His work remains highly controversial though surprisingly influential, perhaps partly because of the circumstances of his death. (He died in 1957 in a US prison during proceedings for contempt of court relating to an injunction brought by the Federal Drug

Administration requiring the destruction of all his accumulators). In 1911 at St. Thomas' Hospital in London the physician Walter Kilner used a glass screen stained with dye to visualise what he called the aura of a human body. He claimed in his 1965 book *The Human Aura* that one could use this image to diagnose physical and mental disorders from the patterns of colours revealed. Similar ghostly images were created by the process of Kirlian photography named after Semyon and Valentina Kirlian who in 1939 developed a method for photographing 'fields' around living things. As well as humans, they captured the spiky corona around animals and plants. Since then subsequent practitioners have photographed the aura of apparently 'inert' objects such as jewellery. This is not merely a diversion among a fringe group of 'New-Agers'. In contemporary scientific research there are a number of well defined strands of activity that take seriously the notion of a vital living energy. Amongst these are the British works of Dr De La Warr and Dr Drown who published respected studies on Radionics and Bio-magnetism in the mid-sixties. In addition, as mentioned, there is a vast range of work in HEF (Human Energy Field) studies covering Light Emission, Electromagnetic Field and Bioplasma studies.

8. All aboard the Holo-deck

Cinema is but one (spectacularly successful) strand in a disparate web of ideas, techniques and devices that sought to capture, store and retrieve the energy that we sense. From the camera obscura to photography to the moving image, with the later addition of sound and colour, each accretion was to mechanically encode a further dimension of perceived reality, including the erotic energy of humans. Given such a well-established, although erratic trajectory it would be reasonable to expect further advances in technology will attempt to encode yet more dimensions of lived experience. Despite the various abortive efforts to represent the sense of smell, the sensations of motion and stereo vision, research into these areas continues, as do efforts to revive the somewhat tired reputation of Virtual Reality and Teledildonics. Given all this, it is within the bounds of our imaginations to foresee a complex of apparatuses that capture and store, perhaps, all human sensory stimuli so that a seamless and transparent evocation of reality is achieved — something along the lines of the 'Holo-deck' in *Star Trek*. Any such device would most likely operate on the same principles as rock painting: that is, arranging matter or energy so as to signify the presence of something that is also absent — to overlay imagination on perception.

9. The Séance: Imagination and perception

It would not be too far-fetched to compare the experience of representation, especially in the cinema, with that of the Séance — both idioms that have pronounced erotic overtones. Bodily forms, objects and scenes are invoked in their absence using techniques of energetic concentration, manipulation and suggestion. Insubstantial apparitions can be summoned at will (given the right conditions) only to disappear once the magic circle of human and operative apparatus is broken (the plug is pulled, the light is turned on). It is this magical illusion that we fall for each time see a picture or read a piece of text. The technique is to trick our organs of perception into perceiving things that are apparently not there in the expectation that the rest of our body will respond as though they are, which it invariably does — hence the sensual response to 'inert' images where the absent becomes present. The efficacy of the medium (spiritual or mechanical) relies on the extent to which this trick succeeds. Vaudeville entertainers such as ventriloquists, impressionists and conjurors operate in an identical way. Imagination, to reiterate, is simply this: the human faculty for perceiving things that are, in one sense, not there but which, in another sense, are. The various tricks and machines we have evolved for the induction of imagined perceptions constitute the technology of representation, from woodcarvings to waxwork museums to AI and VR.

ALBUTT, C. 1895.Nervous Diseases of Modern Life. *Contemporary Review*, (3). pp. 210-231.

AUMONT, J. 1997. *The Image*. London: BFI.

BABBAGE, C. 1961. *Charles Babbage and his Calculating Engines:* Selected Writings by Charles Babbage and Others, Eds. Philip Morrison and Emily Morrison. New York: Dover

BARTHES, R. 1967. (tr. Lavers and Smith) *Elements of Semiology*. London: Jonathan Cape.

BATCHEN, G. 1991. Seeing Things, Vision and Modernity. *Afterimage*, September. pp. 5-7.

BERGSON, H. 1998 (1907). *L'Evolution Créatrice*. New York: Dover.

BIBEL, W., AND JARRAND, P., eds.1987. *Fundamentals of Artificial Intelligence*. New York: Springer Verlag.

BIJKER, W. 1987. The Social Construction of Bakelite: Toward a Theory of Invention. In: Bijker, W. and Law, J. 1992. *Shaping Technology — Building Society: Studies in Sociotechnical Change*. Cambridge, Mass.: MIT Press.

BOLTER, J. 1984. *Turing's Man, Western Culture in the Computer Age*. New York: Penguin.

BOWIE. H.P. 1911. *On the Laws of Japanese Painting*. San Francisco: Elder & Co.

BRASSAÏ. 1999. *Conversations with Picasso*. Chicago: University of Chicago Press.

BRAUN, M. 1993. *Picturing Time, The Work of Etienne Jules Marey* (1830-1904). Chicago: University of Chicago Press.

BRETON, A. 1987 (1937). *L'Amour fou*. Nebraska: University of Nebraska Press.

BUCK-MORSS, S. 1989 *The Dialectics of Seeing: Walter Benjamin and the Arcades Project*. Cambridge, Mass.: MIT Press.

CARLSON, B., AND GORMAN, M. 1990. Understanding Invention as a Cognitive Process: the Case of Thomas Edison and Early Motion Pictures, 1888-91. *Social Studies of Science*, (20). pp. 387-430.

COON, D. 1993. Standardizing the Subject: Experimental Psychologists, Introspection, and the Quest for a Technoscientific Ideal. *Technology and Culture*, (19). pp.757-783.

COOTER, R., AND PUMFREY, S. 1994. Separate Spheres and Public Places: Reflections on the History of Science Popularisation and Science in Popular Culture. *History of Science*, (xxxii). pp. 237-267.

CROWLEY, A. 1991 (1929). *Magick in Theory and Practice*, Secaucus, NJ: Castle Books.

CUTCLIFFE, S., AND GOLDMAN, L., eds.1996. New Worlds, New Technologies, *New Issues*. London: AUP.

DANESI, M. 1990. Thinking, is Seeing: Visual Metaphors and the Nature of Abstract Thought. *Semiotica*, (3/4). pp. 221-237.

DAWKINS, R. 1986. *The Blind Watchmaker*. London: Longman.

DAVIS, M. 1998. *The Ecology of Fear: Los Angeles and the Imagination of Disaster*. London: Picador.

DE LA WARR. G. W. 1967. *Biomagnetism*. Oxford: Delawarr Laboratories

DENNETT, D. 1993. *Consciousness Explained*. London: Penguin.

DENNETT, D. 1984. Computer models and the Mind-a view from the East Pole. *TLS*, December 14. pp. 1453-1454.

DESCARTES. R. 1912 (1637). *A Discourse On Method: Meditations and Principles*. London: Everyman

DRINKA, G. 1984. *The Birth of Neurosis: Myth, Malady, and the Victorians*. New York: Simon and Schuster.

FERGUSON, E. 1992. *Engineering in the Mind's Eye*. Cambridge, Mass.: MIT Press.

FLAUBERT, G. 1989. *Madame Bovary*. London: Penguin

FORSTER, E. M. 1997 (1909). *The Machine Stops (And other Stories)*. London. Deutsch

FOUCAULT, M. 1985. *The History of Sexuality*. Vol.1. London: Penguin.

FRAZER, J. 1998 (1890). *The Golden Bough*. Oxford: Oxford Paperbacks.

FREUD, S. 1985 (1924). *'A Note Upon the 'Mystic Writing-Pad'*. London: Penguin.

FREUD, S. 1985 (1917). *On Metapsychology: The Theory of Psychoanalysis*. London: Penguin.

GOLDSTEIN, J. 1991. The Uses of Male Hysteria: Medical and Literary Discourse in Nineteenth-Century France. *Representations*, (34). pp. 134-165.

GOMBRICH, E. 1999. *The Uses of Images*. London: Phaidon.

GOODING, D, PINCH, T., AND SCHAFFER, S. 1989. *The Uses of Experiment*. Cambridge: Cambridge University Press.

GRAU, R. 1944. *The Theatre of Science*. New York: Broadway Publishing Company.

HACKING, I. 1983. Representing and Intervening: *Introductory Topics in the Philosophy of Natural Science*. Cambridge: Cambridge University Press.

HAUSER, A. 1989 (1951). *The Social History of Art*. London: Routledge.

HAWKING, S. 1988. *A Brief History of Time*. London: Bantam Press.

HAYES, D. 1992. The Growing Inaccessibility of Science. *Nature*, 356 (6372). pp.739-740.

HAYLES, N. K. 1999. *How We Became Posthuman*. Chicago: University of Chicago Press.

HOLMES, F. 1992. Do We Understand Historically how Experimental Knowledge is Acquired. *History of Science*, 30 (88). pp. 119-135.

HOUSE OF LORDS. 2000. *Science and Technology Select Committee Report*. London: HM Stationers.

HUGHES, T. 1986. *American Genesis: A Century of Invention and Technological Enthusiasm,*

1870-1970. New York: Viking.

HUGHES, T. 1983. *Networks of Power: Electrification in Western Society, 1880-1930*. Baltimore: John Hopkins University Press.

HUGHES, H. 1979. *Consciousness and Society: The Reorientation of European Social Thought* 1890-1930. Brighton: Harvester Press.

IHDE, D. 1992. *Instrumental Realism. Bloomington*: Indiana University Press.

JASTROW, J. 1962. *Error and Eccentricity in Human Belief.* London: Dover.

KILNER, W. J. 1965 (1911) *The Human Atmosphere; or the Aura made Visible by the Aid of Chemical Screens (reprinted as The Human Aura)*. New York: Rebman Co.

KIRLIAN S. D. & V. H. 1968. *The Significance of Electricity in the Gaseous Nourishment Mechanism of Plants, Bioenergic Questions – and Some Answers*. U.S.S.R.: Alma Ata.

KITTLER, F. 1990. *Discourse Networks 1800/1900*. Stanford: Stanford University Press.

KITTLER, F. 1980. Gramophone, Film, Typewriter. *October*, (41). pp. 101-118.

KOYRÉ, A. 1957. *From Closed World to the Infinite Universe*. Baltimore: John Hopkins University Press.

LACAN, J. 1977. *Écrits: A Selection*. London: Tavistock.

LATOUR, B. 1987. *Science in Action*. Cambridge, Mass.: MIT Press.

LATOUR, B. 1988. Drawing Things Together. In: M. Lynch and S. Woolgar. *Representation in Scientific Practice*. Cambridge, Mass.: MIT Press.

LAYTON, E. Mirror Image Twins: the Communities of Science and Technology in *19th Century America. Technology and Culture*, 4 (12). pp. 562-580.

LAVERY, HAGUE AND CARTWRIGHT. 1996. *Deny All Knowledge: Reading the X-Files*. London: Faber and Faber.

LENIN, V. 1977 (1908). *Materialism and Empirio-Criticism*. Moscow: Progress Publishers.

LEVY, S. 1992. *Artificial Life*. London: Jonathan Cape

MATLOCK, J. 1991. Doubling out of the Crazy House: Gender, Autobiography, and the Insane Asylum System in *Nineteenth Century France. Representations,* (34). pp. 167-194.

MARX, L. 1988. *The Pilot and the Passenger: Essays on Literature, Technology, and Culture in the United States*. Oxford: Oxford University Press.

MAZLISH, B., ed.1965. The Railroad and the Space Programme An Exploration in *Historical Analogy*. Cambridge, Mass.: MIT Press.

MICHALE, M. 1993. On the Disappearance of Hysteria. *ISIS*, (84). pp. 496-526.

MORUS, I. 1993. Currents from the Underworld, Electricity and Technology of Display in Early Victorian England. *ISIS*, (84). pp. 50-69.

NABOKOV, V. 1950. *Lolita*. London: Weidenfeld and Nicolson

NASAW, D. 1993. *Going Out: The Rise and Fall of Public Amusements*. New York: Basic Books.

NELSON, T. 1974. *Computer Lib / Dream Machines. Redmond Washington*: Tempus Books of Microsoft Press

NIETZSCHE, F. 1994 (1878). *Human, All Too Human*. London : Penguin

NIN, A. 1979. *Delta of Venus*. Florida: Harvest/HBJ

NYE, D. 1990. *Electrifying America, Social Meanings of A New Technology* 1880-1940. Cambridge, Mass.: MIT Press.

NYE, D. *American Technological Sublime*. Cambridge, Mass.: MIT Press.

PAUCALDI, G. 1990. Electricity and Life: Volta's Path To The Battery. *Historical Studies in the Physical and Biological Sciences*, (21). pp. 123-160.

PEPPERELL, R. 1995. *The Post-Human Condition*. Oxford: Intellect.

PUNT, M. 2000. *Early Cinema and the Technological Imaginary. Amsterdam*: Postdigital Press.

REICH, W. 1942. *The Discovery of the Orgone*. New York: Orgone Institute Press.

ROMANYSHYN, R. 1989. *Technology as Symptom and Dream*. London: Routledge.

RUBY, J. 1995. *Secure the Shadow, Death and Photography in America*. Cambridge, Mass.: MIT Press.

SCHAFFER, S. 1992. Self Evidence. *Critical Inquiry*, (18). pp. 327-362.

SCHIFFER, M. 1993. Cultural Imperatives and Product Development: The Case of the Shirt Pocket Radio. *Technology and Culture*, 34 (1). pp. 98-114.

SCHIVELBUSCH, W. 1988. *Disenchanted Night, the industrialisation of Light*. Berkeley: University of California Press.

SCHLUPMANN, H.1996. Cinema as Anti-Theatre: Actresses and Female Audiences in Wilhemianian Germany. In: R. Abel, ed. *Silent Film*. New Brunswick: Rutgers University Press.

SEARLE, J. 1980. Las Meninas and The Paradox of Pictorial representation. *Critical Inquiry*, (6). pp. 477-488.

SHAPIN, S. 1994. *A Social History of Truth: Civility and Science in 17th Century England*. Chicago: University of Chicago Press.

SHAPIN, S. 1996. *The Scientific Revolution*. Chicago: University of Chicago Press.

SHAPIN, S. and SCHAFFER, S. 1985. *Leviathan and the Air Pump: Hobbes, Boyle, and the Experimental Life*. Princeton, NJ: Princeton University Press.

SHELLEY, M. 1968 (1818). *Frankenstein*. USA: Minster Classics

SILVERMAN, R. 1993. The Stereoscope and Photographic Depiction in the 19th

Century. *History of Technology* Fall. pp.729-756.

SMITH, M., AND MARX, L., eds. 1996. *Does Technology Drive History? The Dilemma of Technological Determinism.* Cambridge, Mass.: MIT Press.

SOBCHACK, V. ed.1996. *The Persistence of History: Cinema Television and the Modern Event.* New York: Routledge.

STAFFORD, B. 1991. *Body Criticism.* Cambridge, Mass.: MIT Press.

STAFFORD, B. 1994. *Artful Science.* Cambridge, Mass.: MIT Press.

SUCHMAN, L. 1987. *Plans and Situated Actions.* Cambridge, Mass.: MIT Press.

TAMULIS, A. et al. 2000. *Design of Basic Elements of Digital and Postdigital Computers Based on Quantum Mechanical Investigation of Fullerene and Photoactive Molecules.* www.foresight.org

THOM, R. 1972. *Structural Stability and Morphogenesis.* London: Addison-Wesley.

TURING, A. 1986. Volume 10 in The Charles Babbage Institute Reprint Series for The *History Of Computing: A. M. Turing's ACE Report of 1946* and other papers Cambridge, Mass.: MIT Press.

TURNER, M. 1993. *Reading Minds: The Study of English in the Age of Cognitive Science.* Princeton, NJ: Princeton University Press.

TWEENY, R. 1992. *Stopping Time: Faraday and the Scientific Creation of Perceptual Order. Physis,* (129). pp. 149-164.

ULMER, G. 1989. *Teletheory: Grammatoloav in the Age of Video.* New York and London: Routledge

WIENER, N. 1948 & 1961. *Cybernetics or Control And Communication in The Animal and The Machine.* Cambridge, Mass.: MIT Press.

WIENER, N. 1985. *Norbert Wiener: Collected Works with Commentaries.* Cambridge, Mass. Press

WIENER, N. 1961. *Cybernetics.* Cambridge, MA: MIT Press.

WILLIAMS, R. 1990. *Notes on the Underground: An Essay on Technology, Society and the Imagination.* Cambridge, MA: MIT Press.

WINKLER, M.G., AND VAN HELDEN, A. 1992. Representing the Heavens, Galileo and Visual Geometry. *ISIS,* (83). pp. 195-217.

WINTER, A. 1998. *Mesmerized, Powers of Mind in Victorian Britain.* Chicago: University of Chicago Press

WINTER, A. 1994. Mesmerism and Popular Culture in Early Victorian England. *History of Science,* xxxii. pp. 317-343.

WITTGENSTEIN, L. 1981 (1953). *Philosophical Investigations.* Oxford: Blackwell.